Cambridge International AS and A Level

Geography WORKBOOK

Geographical skills

Garrett Nagle
Paul Guinness

To Angela, Rosie, Patrick and Bethany

The Publishers would like to thank the following for permission to reproduce copyright material:

Artwork: p.11 Cartoon © Louis Hellman; **p.19** Skinner, B. and Porter, S., Variations in discharge, sediment load and sediment composition along the River Nile and its main tributaries. Adapted from *The Dynamic Earth* (Wiley-Blackwell 1995); **p.22** Box and whiskers diagram to show the impact of selected activities on warming (positive values) or cooling (negative values), adapted from *World Development Report, 2010: Development and climate change* (The World Bank, 2010). Licensed under Creative Commons Attribution license CC BY 3.0 IGO (http://creativecommons.org/licenses/by/3.0/igo/); **p.23** Hofer, T. and Messerli, B., Average monthly rainfall in the Ganga–Brahmaputra–Meghna Basin, from *Floods in Bangladesh: History, Dynamics and Rethinking the Role of the Himalayas* (United Nations University Press/FAO 2006); **p.43** United Nations Department of Economic and Social Affairs/Population Division, Age of international migrants in LICs/MICs and HICs, adapted from *International Migration Report 2013* (United Nations, 2013). Reprinted with the permission of the United Nations; **p.47** *left* GeoDiver/City – UMR Geographie-cites CNRS, Semi-log graph: population of country/region as a percentage of world urban population; *right* GeoDiver/City – UMR Geographie-cites CNRS, Double log (log–log) graph: city size distributions in seven countries around 2010; **p.56** *top* Department of Rural Development and Land Reform (South Africa), The Cape Peninsula, South Africa; **p.73** Distribution of proven oil reserves, 1994, 2004, 2014. Adapted from *BP Statistical Review of World Energy 2015*. Reproduced by permission of BP p.l.c. (www.bp.com/statisticalreview); **p.77** World Trade Organization, Scatter graph with proportional circles, adapted from *International Trade Statistics, 2014* (www.wto.org/english/res_e/statis_e/its2014_e/its2014_e.pdf). **Tables: p.14** *middle* Simmons, I., Land subsidence and groundwater extraction, from *Changing the Face of the Earth* (Wiley-Blackwell, 1989); **p.16** Hofer, T. and Messerli, B., Percentage of Bangladesh covered by floods, 1954–2004, from *Floods in Bangladesh: History, Dynamics and Rethinking the Role of the Himalayas* (United Nations University Press/FAO 2006); **p.33** Carl Haub and Toshiko Kaneda, Demographic data for six countries, Selected data, 2014, from *2014 World Population Data Sheet*, Population Reference Bureau; **p.52** The World Bank, The trend in global ecosystem services, from *World Development Report 2010: Development and Climate Change* (The World Bank, 2010). Licensed under Creative Commons Attribution license (CC BY 3.0 IGO); **p.63** Armstrong, D. et al., Top twelve most deadly volcanic eruptions, from *Geology* (Heinemann/Pearson Education Limited 2008). **Text: p.15** Hofer, T. and Messerli, B., Flooding in Bangladesh, from *Floods in Bangladesh: History, Dynamics and Rethinking the Role of the Himalayas* (United Nations University Press/FAO 2006).

Every effort has been made to trace all copyright holders, but if any have been inadvertently overlooked the Publishers will be pleased to make the necessary arrangements at the first opportunity.

Orders: please contact Hachette UK Distribution, Hely Hutchinson Centre, Milton Road, Didcot, Oxfordshire, OX11 7HH. Telephone: +(44) 01235 827827. Email education@hachette.co.uk. Lines are open from 9 a.m. to 5 p.m., Monday to Saturday, with a 24-hour message answering service. You can also order through our website: www.hoddereducation.com.

ISBN: 978 1 4718 7376 8

First published in 2016 by

Hodder Education,

An Hachette UK Company

Carmelite House

50 Victoria Embankment

London EC4Y 0DZ

www.hoddereducation.com

Impression number 10 9 8 7 6 5 4

Year 2021

Illustrations by Aptara Inc.

Typeset in Caecilia LT Std, 10/13 pts. by Aptara Inc.

Printed in Great Britain by Ashford Colour Press Ltd.

A catalogue record for this title is available from the British Library.

Contents

Geographical skills

As part of your Cambridge International AS and A Level Geography course, you need to be familiar with a variety of source materials, and these should be integrated into your course. This Geographical Skills Workbook will help you to practise a variety of skills with reference to geographical content that you have studied. Many of the skills tested here use a case-study approach that provides useful information to help build up your knowledge and understanding of Geography. The list of skills/resources below is not exhaustive but gives a general idea of the range of material that you need to be familiar with:

Graphs	Bar charts, divided bar charts, line graphs, scatter graphs (including line of best fit), pie charts, proportional circles, triangular graphs, climate graphs, etc.
Photographs	Colour, black/white, aerial, terrestrial, satellite
Maps	Survey maps (1:25 000 and 1:50 000 scales), flow line, isoline, choropleth, sketch, etc.
Diagrams	Two and three dimensional, with/without annotation, flow diagrams, etc.
Written	Text from a variety of sources (including newspapers, articles, books, interviews)
Numeric	Tables, charts, raw data, etc.
Cartoons	

Graphs

☐ Bar charts

In bar charts, the length of each bar represents the *quantity* of each component; that is, places or time intervals. The vertical axis has a scale that measures the total of each of these components. There are five main types of bar chart, as follows:

- **Simple bar chart** – each bar indicates a single factor (see for example Figure 1.9 on page 6 of the *Cambridge International A and AS Level Geography* textbook [second edition])
- **Multiple or group bar chart** – features are grouped together on one graph to help comparison (see for example Figure 5.25 on page 137 of the textbook)
- **Divided (compound) bar chart** – various elements or factors are grouped together on one bar (the most stable element or factor is placed at the bottom) (see for example Figure 4.24 on page 99 of the textbook)
- **Percentage divided (compound) bar chart** – a variation on the compound bar chart, used to compare features by showing the percentage contribution; these graphs do not give a total in each category, but compare relative changes in percentages (see for example Figure 4.11 on page 90 of the textbook)
- **Median line bar chart** – a line is added to show the median value (the middle value after data has been ranked).

☐ Age/sex structure diagram

One widely used variant of the multiple bar chart is the age/sex structure diagram (see for example Figure 4.14 on page 91 of the textbook). This is constructed in 1, 5 or 10-year age groups, with males on the left-hand side and females on the right-hand side. The standard age/sex structure diagram uses 5-year groups. This almost always takes the form of a pyramid, with the youngest age group at the base and the oldest at the top. The horizontal bars are drawn proportional in length to either the percentage of the total population or the actual number in each group.

☐ Line graphs

Line graphs are relatively simple graphs that show change over time (see for example Figure 1.29 on page 20 of the textbook). A line graph that shows population change for 2010–16 can show change between 2010 and 2011, 2011 and 2012, and so on. Line graphs use *continuous data* and they show *trends*. These trends can be *absolute* or they can be *relative*. Line graphs can be *simple* – representing one feature – or they can be *multiple* – showing many features (see for example Figure 1.3 on page 2 of the textbook). In all line graphs, there are *independent* and *dependent variables*. The independent is plotted on the horizontal or x-axis and the dependent on the vertical or y-axis.

☐ Scatter graphs

Scatter graphs show how two sets of data are related to each other, for example population size and number of services, or distance from the source of a river and average pebble size. To plot a scatter graph, it must be decided which variable is independent (in the latter example, organic content or distance from the source) and which is dependent (number of services or average pebble size). The independent is plotted on the horizontal or x-axis and the dependent on the vertical or y-axis. For each set of data, a line is projected from the corresponding x- and y-axes, and where the two lines meet a dot or cross is marked. A line of best fit is added if there is a relationship between the two variables. This allows anomalies to be identified.

☐ Pie charts

Pie charts are frequently used on maps or on their own to show variations in size and composition of a geographical feature (see for example Figure 10.20 on page 331 of the textbook). Every 3.6° on the pie chart represents 1 per cent; thus, the 360° of the circle represents 100 per cent. To plot values, they are first converted to percentages and then multiplied by 3.6. This gives the number of degrees of each segment.

☐ Proportional circles

Proportional circles are the next step up from pie charts (see for example Figure 6.29 on page 169 of the textbook). While pie charts are viewed as a basic graphical technique, proportional circles are a higher-level technique. Proportional circles are useful when illustrating the differences between two or more amounts. They are particularly effective when placed on location maps. It is most desirable to place the symbols inside the boundaries of the regions the values refer to. However, if some symbols cannot fit comfortably within their regions, they are placed in the nearest comfortable space and *leader lines* are used to connect them with their regions.

The area of a circle is found by using the formula πr^2; therefore, the circles are drawn in relation to the square root of the value. The size of the largest circle to be used is decided and its radius (r) written down. The square roots of all the values to be mapped are worked out. Using the square root of the largest value to be mapped, the value (v) that it must be multiplied by in order to make the actual circle on the map is worked out, then all the other square roots are multiplied by the value (v).

The advantages of pie charts and proportional pie charts are that they are very clear, can show a large amount of data, and can be very striking. The main disadvantage – especially of proportional pie charts – is that they may over-emphasise large values, and so small values are not as clear. They also require time, care and patience to draw.

☐ Proportional semi-circles

Proportional semi-circles can be used to show two sets of data, for example differences between two years. When carefully constructed with contrasting colouring/shading and a clear key, this can be a good technique, particularly when there is a lot of data to illustrate.

☐ Proportional squares, triangles and bars

Proportional squares are constructed in a similar way to proportional circles. Here, the square root value is used to decide the length of the side of the square. Proportional squares are positioned in as central a position as possible over the area they represent. If two squares overlap, it is possible to lighten the shading of one square for the purposes of clarity. In some cases, proportional triangles are used (see for example Figure 10.10 on page 318 of the textbook). Like circles, proportional squares can be sub-divided.

In proportional bars, the length of a bar is proportional to the value it represents. Again, bars can be divided to illustrate a number of different categories.

☐ Triangular graphs

Triangular graphs are used to show data that can be divided into three parts, for example soil (sand, silt and clay), employment (primary, secondary and tertiary) and population (young, adult and elderly) (see for example Figure 3.17 on page 73 of the textbook). It requires the data in the form of a percentage, and to total 100 per cent. The main advantages of triangular graphs are that:

- they allow a large amount of data to be shown on one graph
- groupings are easily recognisable – in the case of soils, groups of soil texture can be identified
- dominant characteristics can easily be shown
- classifications can be drawn up.

They can be tricky and it is easy to get confused, especially if care is not taken. However, they provide a fast, reliable way of classifying large amounts of data that have three components.

☐ Climate graphs

Climate graphs or climographs describe the *seasonal* pattern of precipitation and temperature (see for example Figure 7.1 on page 187 of the textbook). A simple climate graph includes a line graph to show variations in temperature, and a bar chart to show variations in rainfall. They show a number of highly useful figures, such as the *mean monthly average temperature* and *monthly rainfall*. A more complex range bar can be used to show the *mean monthly maximum* and the *mean monthly minimum*. The mean monthly maximum is the average of all the maximum temperatures for each day of the month and the mean monthly minimum is the average of all the minimum temperatures recorded for that month. Rainfall is generally shown as a series of bars. Different scales are shown on the diagram: in the textbook example, the temperature scale is shown by a range bar indicating mean monthly maximum and minimum, and a dot showing the mean monthly average temperature. Rainfall is also shown by bars but has a separate scale, underneath that for temperature.

Reading the climograph

Look out for:

- total rainfall
- seasonality (when most of the rain occurs)
- maximum temperature
- minimum temperature
- range of temperature (maximum to minimum)
- length of time (if any) below freezing.

☐ Dispersion diagrams

A dispersion diagram is a very useful diagram for showing the range of a set of data and their tendency to *group* or *disperse*, and also for comparing two groups of data. It involves plotting the values of a single variable on a vertical axis. Technically, there is a short horizontal axis showing frequency. What is revealed is the frequency distribution. Dispersion diagrams can be used to determine the *class intervals* for choropleth maps.

A critical part of analysing any array of values by means of the dispersion diagram is to determine the *median*, *upper* and *lower quartile* values and therefore the *inter-quartile range*.

☐ Polar graphs and rose diagrams

A polar graph is used to show *direction* as well as *magnitude*, whereas a rose diagram is used to show the composition of a feature.

The method for constructing a polar graph is quite simple. Using a compass and a protractor, lines are drawn that correspond with north (0°), north east (45°), east (90°), south east (135°), south (180°), south west (225°), west (270°) and north west (315°).

For a rose diagram, the compass points (or other fractions, such as sixths) become different component parts of a whole, for example the environmental footprint of different cities.

☐ Kite diagrams

A kite diagram is a form of line graph where the scale is split in two; that is, half the value is shown above a horizontal line and half below. They are most commonly used to show vegetation distribution as, for example, across a sand-dune system. The section for the sand-dune profile is an example of a cross section.

☐ Semi-log and double-log (log–log) graphs

Semi-logarithmic graph paper has a 'normal' scale on one axis – usually the horizontal or y-axis – and a *logarithmic scale* on the other. Semi-log graphs are used to show data that have a particularly wide range of values (see for example Figure 1.25 on page 17 of the textbook).

The logarithmic scale compresses the range of values. It gives more space to the smaller values and reduces the amount of space for the larger values; thus, it can show relative growth quite clearly. On the scale, there are *'cycles' of values*. Each cycle increases by a larger amount, usually to the power of 10.

In some cases, logarithmic scales can be used on both axes. This is known as a **double-log scale**. The Hjülstrom curve is drawn on a log–log scale (see for example Figure 1.13 on page 10 of the textbook). This is used when both sets of data have large ranges.

Photographs

Many photographs that are used in exams are aerial views that show industrial, residential, recreation and commercial land uses (see for example Figure 12.12 on page 387 of the textbook).

Maps

☐ Grid and square references

Grid references are the six-figure references that locate precise positions on a map.

- The first three figures are the *eastings* and these tell you how far a position is across a map.
- The last three figures are the *northings* and these tell you how far up the map a position is.

An easy way to remember is 'along the corridor and up the stairs'; that is, along (eastings first) and then up (or northings). Sometimes, a feature covers an area rather than a point, for example a lake covers a large area rather than a specific point. A grid reference does not tell us about the area of the lake, although it could tell us about its centre. Instead, we use four-figure square references:

- The first two numbers refer to the *eastings* – in particular, the eastings on the left of the square.
- The second two numbers refer to the *northings* – in particular, the northings at the bottom of the square.

☐ Magnetic declination

Direction on a map is generally expressed in one of two ways:

- compass points, for example south west
- compass bearings or angular direction, for example 45°.

Sixteen compass points are commonly used. Compass bearings are more accurate than compass points but can be quite confusing. Compass bearings show variation from magnetic north (magnetic north pole).

- **true north** is the direction of the North Pole
- **grid north** is the parallel lines on an Ordnance Survey map running from north to south
- **magnetic north** is the direction from the magnetic north pole.

The magnetic north pole is located away from the true North Pole, so compasses do not point to true north. In addition, the magnetic north pole is moving – very slowly, but it means that compass bearings will differ each year; this is known as **magnetic declination**.

☐ Flow lines, desire lines and trip lines

Flow maps are used to show the movement of goods, traffic, people and information between places. Flow lines, desire lines and trip lines are all used to illustrate the volume and direction of movement from place to place. The most basic of these techniques is the use of **trip lines**, which use uniformly thin and straight lines to link points of origin and destination (see for example Figure 1.2 on page 1 of the textbook). Trip lines might be drawn to show where shoppers live in a market town; such a trip-line map might link 20 villages in the surrounding area to the market town, each link being shown by a thin straight line. Invariably some lines will be short and others much longer in length.

In contrast to trip lines, flow-line and desire-line diagrams use lines that are proportional to the volume of movement. The difference between the last two is that **flow lines** follow actual routes, while **desire lines** are drawn directly from the point of origin to the point of destination. Desire lines are particularly effective in showing direction of movement and can be impressive when superimposed on a base map. Desire lines are often used to show:

- migrations of populations both within and between countries
- trade between countries
- tourism
- water flow.

Flow lines might be used to show the volumes of traffic from different smaller settlements into a larger settlement, variations in volume of traffic within an urban area or the volume of nutrients being recycled (see for example Figure 7.38 on page 212 of the textbook). Thus, a line 10 millimetres wide may represent 500 vehicles an hour along a road. On the same scale, a line 2 millimetres wide would represent 100 vehicles an hour. Directions of movement are indicated by arrows. Flow lines could also be used to show the number of buses coming into a town on a particular day. The success of using proportional flow lines depends very much on an appropriate choice of scale.

☐ Choropleths

A choropleth is a map that shows *relative density per area* (see for example Figure 9.41 on page 297 of the textbook). Areas are coloured or shaded according to a prearranged key, with each colour or shading representing a range of values. Choropleth maps can use variations in colour or different densities of black and white shading. This technique is good for showing comparisons between places. The base maps for choropleth maps show regions or areas, such as the boroughs in London or the counties in England. The steps below are followed in the construction of a choropleth map:

- A good, clear base map is found.
- The range of data is divided into classes. There should be no fewer than four classes and no more than eight. In the former case, there would not be enough variation on the map; and in the latter situation, the map would contain so much information that it could be confusing.
- The class or category values do not overlap so, for example, we may have 0–19.9, 20–39.9, 40–59.9.
- A colour or density shading is allocated to each class. The convention is that shading gets darker as values increase.
- Each colour is applied to the relevant areas of the map.
- A key, scale and north point are provided.

The choropleth is a popular technique, frequently used in atlases, textbooks and many other types of publication. It can convey a lot of information in a straightforward and visually appealing way. The main disadvantage of the choropleth is that it can show abrupt changes at boundary lines, when in reality change is much more gradual. It also gives the impression of uniformity within individual areas on the map, when in reality a reasonable degree of variation may be present. Careful selection of class sizes can reduce this problem.

☐ Isolines

Isolines join points of equal value on a map (see for example Figure 2.42 on page 56 of the textbook). They are similar to the contours on an Ordnance Survey map that join points of equal height. Isolines can only be drawn when the values under consideration change in a fairly gradual way over the area of the map. Data for quite a large number of locations is required in order to draw a good isoline map. If the data available is patchy or the spatial distribution is complex and disjointed, this technique is difficult to apply as too much guesswork is involved.

Usually a fixed interval is selected, such as the contour interval on an Ordnance Survey map, so that the reader can sense how quickly a phenomenon changes even without looking closely at the figures. An isoline passes between values that are higher and lower so that all values on one side of the line will be higher, and all those on the other side will be lower. The space between different value isolines can be coloured or shaded in the same way that relief is illustrated on atlas maps. If colouring or shading is used then a key should be included.

Isolines are typically used to represent:

- air pressure in the form of isobars
- temperature in the form of isotherms
- rainfall in the form of isohyets
- lines of equal travel time in the form of isochrones
- noise contours around an airport
- pedestrian densities in a CBD
- variations in house prices in urban areas.

☐ Dot maps

Dot maps are used to illustrate the distribution of phenomena such as population, houses or plants by creating a visual impression of density (see for example Figure 11.41 on page 362 of the textbook). The pattern of dots can be analysed statistically, for example by nearest neighbour analysis.

Two parameters must initially be considered: the graphical size of each dot and the value associated with each dot. The dot value should be carefully chosen so that it is high enough to avoid excessive overcrowding of dots in areas with a high concentration of the variable, but low enough to prevent areas with low concentrations having no dots at all, which could give a false impression of emptiness. Ideally, dots should not merge. Dots are positioned on the map to show as closely as possible the distribution of the variable being mapped.

The main advantage of using a dot map is that it can be very effective in providing a good visual impression of variations in spatial distributions, providing the dots are carefully positioned. To do this, detailed information is needed before a map of this kind is constructed. However, large numbers of dots are difficult to count and thus it can be difficult to estimate actual figures in terms of, for example, the population of a densely populated area.

Diagrams

Diagrams include field sketches and sketches based on photographs. A field sketch is a hand-drawn summary of an environment. The sketch should highlight important geographical features. Similarly, a sketch based on a photograph should illustrate the most important features shown.

Written

☐ Writing frames

There are a number of writing frames for answering an essay or a report. In general, the essay title and the material to be included will suggest what type of structure should be used. However, for all questions you need to:

- examine closely the wording of the question
- plan your answer.

It is better to spend time thinking and planning, so that you do not waste time writing about irrelevant material: 35 minutes of relevant material is better than 45 minutes of irrelevant material.

☐ Reading the question

Read the question carefully and underline the command words and the topic to be discussed. There may be some words such as 'and' or 'either ... or'. Questions with 'and' in them generally ask for some factual comment and then require some interpretation. Often the interpretation is more important than the recall of fact. Questions stating 'with the use of examples ...' may allocate one-third or half of the marks for the examples used. If you do not answer the question you cannot get the marks.

☐ Command words

Command words tell you what to do in the essay or how to use the material. There are a number of command words, including:

Command word	What it means
Account	Give reasons for
Assess	Make an informed judgement based on evidence
Calculate	Work out a numerical answer; in general, working should be shown, especially where two or more steps are involved
Compare	Describe both similarities and differences of things; two separate descriptions do not make a comparison
Contrast	Describe differences between two things
Define	State the precise meaning of a term, idea or concept
Describe	State in words the key characteristics and give factual details
Discuss	Present points for and against, or present different viewpoints
Draw	Make a sketch or simple, freehand drawing; may be used with labels
Evaluate	Make a judgement from available evidence
Examine	Investigate closely (describe, explain, offer evidence and comment)
Explain	Set out reasons, causes or purposes
Give	Provide an answer from, or in relation to, a resource
Give reasons	Provide points of explanation
Give the meaning of	State the definition of a term, idea or concept
How	Describe in what way(s) or by what means
How far do you agree	Make an informed assessment, based on evidence
Identify	Name or select one or more characteristics
Label	Add specific names or details to a diagram, graph or map
Name	Provide the appropriate name or term
Outline	Set out the main characteristics, restricted to giving essentials, without supporting details
State	Give a concise answer with little or no supporting argument, expressed in clear terms
Suggest	Apply knowledge and understanding to an unfamiliar situation where there is no single correct answer
To what extent	Form and express a judgement after examining evidence
What	Provide specific information
Which	Provide specific information
Why	Explain the reason or purpose

☐ Types of essay

There are three main generic types of essay: description, explanation and evaluation, although not all essays fit into this classification. **Descriptive** essays are the easiest and require factual recall. Very few, if any, A2 essays will be purely descriptive. By contrast, **explanation** requires you to give reasons and account for why a particular object is the way it is. Lastly, **evaluation** expects an opinion based on the evidence presented throughout the essay. Alternatively, you may be given a ready-made evaluation and asked to say how far you agree.

Writing an essay

It is vital to think about the essay and to plan it. Quality is more important than length. One way of planning an essay is known as the *points–group–order* method. Write down a list of points that are relevant to the essay and then group them, and finally put them into order of importance.

Mind maps or branching maps are a useful way of selecting and presenting information. There is more than one way of answering most essays; so as long as the structure is logical and clear, and it answers the questions, it is acceptable. To find a structure requires a general overview of the subject as well as the small details to support your views.

The introduction

Writing the introduction is a key skill. Most examiners have a good idea of the grade a candidate will achieve after they have read the introduction! The introduction should:

- define the terms used in the title
- show the line of argument that will be taken
- list the factors that are important
- state which examples and case studies will be used.

The introduction needs to be clear and full of impact, rather like the introduction in a newspaper article that catches your attention, shows the main line of argument and leaves you wanting to read more.

The body

The main body of the essay develops the arguments. Each paragraph should have a key sentence or key point and the rest of the paragraph explains and provides evidence. Paragraphs must be linked, and this is done in a variety of ways, for example by:

- referring back to the point above ('the result of this is to cause ...')
- linking in a time sequence ('next, after, and so on')
- comparing ('there are also environmental problems in shanty towns')
- contrasting ('by contrast, the China economy is based on export-led manufacturing').

Throughout the essay, the quality of language needs to be high; thus, it is important to use key words and phrases that make the essay read well, as well as allowing it to flow. These words and phrases are small but have a major impact on how clearly the arguments come across.

The conclusion

The conclusion is more than just a summary. It may:

- assess the changing nature of the topic
- examine the changing importance of factors involved
- draw out the uniqueness of the material used (every example is different)
- look at the contrasts between HICs and MICs/LICs
- look to the future (how will the subject change in the next 25 years?)
- end with a question, for example 'Even if we can predict earthquakes and volcanoes, can we stop people from living in hazardous areas?'.

☐ Key words and phrases in Geography essays

Adding	Drawing conclusions	Comparing and giving contrasting examples
and in addition moreover furthermore indeed again similarly likewise also next first … last finally subsequently too what is more as well as at this stage at the same time on the one hand	so thus therefore then hence generally consequently because since to this extent at most in conclusion as a result because of this in this way due to in the case of nevertheless	thus however whereas although on the other hand by contrast an exception to this even though instead of in spite of for example when in particular for instance such as

Numeric

There are different types of numeric data. The simplest data – *nominal* – can yield only simple, albeit important, information, whereas more complex data can be tested with more complex techniques.

☐ Qualitative and quantitative analysis

Qualitative analysis occurs when the type of measurement is nominal (named) or ordinal (ranked). In contrast, **quantitative analysis** occurs when the data is numeric, interval or ratio.

Nominal data

Nominal data refer to objects that have names (for example rock types, land uses, dates of floods). Nominal data are the most basic type of data. In a geographical investigation, we might categorise settlements as hamlets, villages or towns, and then count them to form a hierarchy; alternatively, when studying urban function, we may classify areas as industrial, commercial, residential, recreational, and so on (these may later be expressed as a percentage of total land use) – this is *nominal data*.

Ordinal data

Ordinal, or ranked, data refer to objects when they are placed in ascending or descending order (for example China, India, USA, Indonesia in terms of population size, or the largest emitter of carbon dioxide, the second largest, and so on). Settlement hierarchies are often expressed in terms of ranks. One of the most commonly used statistics in Geography is the Spearman's Rank Correlation Coefficient, which compares two sets of ranked data, for example gross national income and infant mortality rates.

Interval and ratio data

Interval data and ratio data refer to real numbers. Interval data is different from ratio data. In interval data, there is no true zero. This means, for example, that if Cairo has an average temperature of 30 °C and London 15 °C, it is not possible to say that Cairo is twice as hot as London. This is because if the Fahrenheit Scale or the Kelvin Scale were used, the numbers would be different. On the other hand, ratio data does possess a true zero. It is possible to say that London's 600 millimetres of rain is approximately 20 times greater than Cairo's 30 millimetres of rain. The ratio is the same if we were to use inches.

☐ Tables

Tables are used to record and present data. Scanning systems are used in quality of life and other types of survey, such as an environmental quality index. It may be quite subjective, but as long as the same standards are kept throughout the survey, it should not affect the results. Scanning systems can be carried out in the field or from a photograph.

☐ Tally charts

Tally charts are simply counts of an element, for example the number of pedestrians at different parts of the CBD.

Cartoons

A cartoon is a diagram that may exaggerate a feature to draw attention to a particular point. What does this cartoon suggest about the homogenisation of urban areas?

Working with maps and satellite images

See the textbook pages 487–89.

1 **Mapwork skills: 1:50000 map extract of the Grand Canyon and Grand Canyon village**

Study the 1:50000 map of part of the Grand Canyon on page 487 of the textbook.

a State the four-figure grid reference for Grand Canyon Village.

____

b In which square are there three golf courses?

____

c Identify the feature at 376464.

____

d In which direction is the Colorado River flowing?

____

e What is the approximate altitude of the Colorado River?

____

f State the approximate altitude of the Sumner Butte in the north-east section of the map extract.

____

g Calculate the approximate distance of Highway 64, as shown on the map extract.

____

h Identify the landform that makes up the southern part of the map extract.

____

i Comment on the contour pattern in squares 3239 and 3743.

____

____

j Describe the contour pattern of a butte, for example Sumner Butte in square 3848.

____

2 **Mapwork skills: 1:25000 map extract of Vancouver**

Study the 1:25000 map of Vancouver on page 488 of the textbook.

Photo 1

Photo 2

Photo 3

a State the four-figure square reference for Beaver Lake, towards the north of the map.

b Name the main park in the northern part of the map extract.

c State the main direction of Nelson Street.

d Suggest why sites were chosen in 2108 and 2208 to build bridges across the bay.

e Photo 1 above shows a view of central Vancouver from Stanley Park. In which direction was the camera pointing?

f Photo 2 shows Vancouver's famous gas clock on Water Street, central Vancouver. In which grid square is Water Street?

g Photo 3 shows Vancouver's port. Comment on the main features of the port, as shown in the photo.

h Comment on the physical geography of Vancouver, as shown in photo 3 and on the map extract.

3 Satellite images: the changing Aral Sea

The satellite images on page 489 of the textbook show the Aral Sea in 1989 and 2014.

a Describe the main changes that have occurred in the Aral Sea, as shown on the satellite images.

b How does the use of satellite images aid our understanding of the changes to the Aral Sea (see the Case Study text and Figure 1.27 on page 19 of the textbook)?

Paper 1 Core Physical Geography

Topic 1 Hydrology and fluvial geomorphology

1 Changes to groundwater

The table below shows the decline in the water table for selected locations.

Table 1.1 Decline in water table (metres/year)

North China Plain	2.3
Guam, Mexico	1.5–3.5
Beijing, China	0.91
Manila, Philippines	9.1
North-east Iran	2–3

a Comment on the drop in the water table as shown in Table 1.1.

Table 1.2 Land subsidence and groundwater extraction

	Maximum subsidence (m)	Area affected (km²)
San Joaquin Valley, California, USA	9	13 500
Houston-Galveston, Texas, USA	2.75	12 170
Eloy-Picacho, Arizona, USA	3.6	8 700
Tokyo, Japan	4.6	2 400
Nobi Plain, Japan	1.5	800
Po Valley, Italy	3	780
Venice, Italy	0.14	400
London, UK	0.35	450
Mexico City, Mexico	8.7	225

Simmons, I., 1989, *Changing the Face of the Earth*, Blackwell, Table 5.32, page 277

b Using Table 1.2, explain the likely impact of extracting water.

2 Flooding in Bangladesh

Read the media headlines below and answer the questions that follow.

- Nepal has lost half its forest cover within a thirty-year period (1950–1980) and by AD2000 no accessible forest will remain. (World Bank, 1979)
- The severity of the recent floods in Bangladesh has led the government to look for a flood plan which would, in the long term, provide a comprehensive and permanent solution to the recent flood problem and so create an environment for sustained economic growth and social improvement (World Bank, 1989)
- Bangladesh in grave danger: deforestation in Himalayas aggravating floods (*Bangladesh Observer*, 2 June 1990)
- When the Himalayas were covered in trees, Bangladesh suffered a major flood about twice a century; one every four years is now the average (UNEP, 1992)
- The severe floods in eastern India and Bangladesh are not the result of a natural disaster, but of a ruthless exploitation of wood which has been practised over centuries in the forests of the Himalayas (*Basler Zeitung*, 15 September 1998)

Hofer, T., and Messerli, B., 2006, *Floods in Bangladesh,* United Nations University Press

a Briefly explain the link between deforestation and flooding.

____

____

____

b Suggest other factors, apart from deforestation, that may be responsible for flooding in Bangladesh.

____

____

____

c Outline the ways in which flooding limits economic growth and social improvement.

____

____

____

d Suggest why the Bangladeshi government may not be able to provide a permanent and complete flood plan for Bangladesh.

____

____

____

e Comment on the accuracy or bias of the newspaper headlines.

____

____

____

____

3 Changes in the annual extent of flooding in Bangladesh, 1954–2004

Study the table of data, which shows the percentage of Bangladesh covered by floods.

Table 1.3 Percentage of Bangladesh covered by floods, 1954–2004 (for years in which there is no value, data were not available)

Year	% of country flooded (by area)	Year	% of country flooded (by area)
1954	25.77	1980	23.11
1955	35.37	1981	
1956	24.79	1982	2.20
1957		1983	7.77
1958		1984	19.75
1959		1985	7.98
1960	19.89	1986	3.22
1961	20.17	1987	40.13
1962	26.05	1988	63.01
1963	30.19	1989	4.27
1964	21.71	1990	2.45
1965	19.89	1991	19.38
1966	23.39	1992	1.36
1967	18.00	1993	19.48
1968	26.05	1994	0.28
1969	29.00	1995	21.68
1970	29.70	1996	24.26
1971	25.42	1997	
1972	14.57	1998	67.93
1973	20.87	1999	22.26
1974	36.84	2000	24.19
1975	11.63	2001	2.71
1976	19.82	2002	10.16
1977	8.75	2003	14.57
1978	7.56	2004	37.27
1979		**Average**	**20.78**

Adapted from Hofer and Messerli, Table 4.1

a Copy this data onto a spreadsheet and plot the data for the annual variation in flood coverage on the grid below.

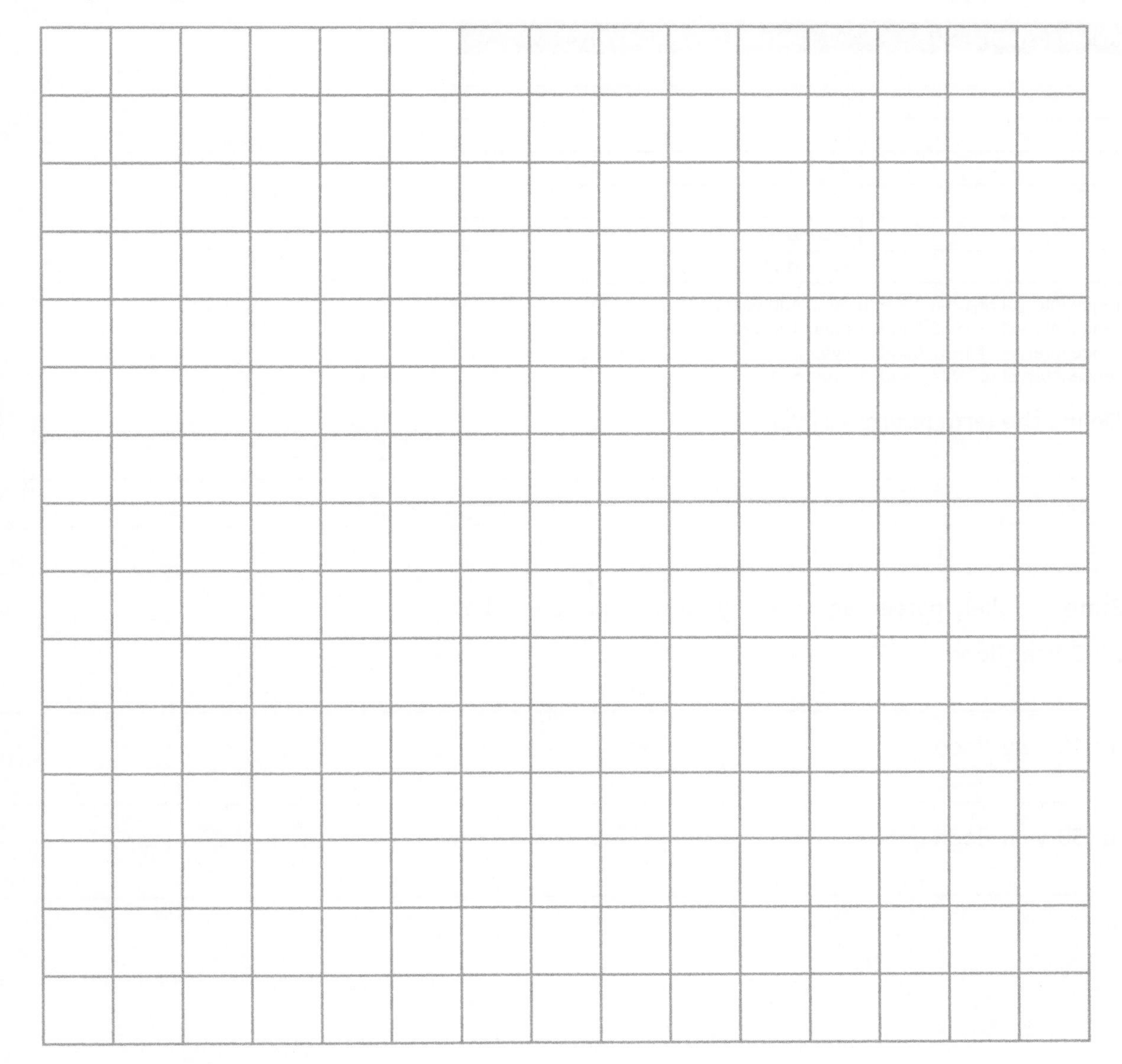

b Identify the five years with the:

i highest coverage

____

ii lowest coverage.

____

c Comment on the annual variability in the extent of flooding in Bangladesh.

____

____

____

____

The table below provides data on the recurrence interval of different floods in Bangladesh.

Table 1.4 Recurrence interval of floods

Return period (years)	Affected area (% of country)
2	20
5	30
10	37
20	43
50	52
100	Around 60
500	Around 70

Recurrence interval of 1998 flood: around 400 years
Recurrence interval of 1988 flood: around 100 years
Recurrence interval of 1987 flood: 20 years
Recurrence interval of 1974 flood: 10 years

d Define the term *recurrence interval.*

e State the likely percentage of Bangladesh to be covered by a:

i 2-year flood

ii 10-year flood

iii 50-year flood.

4 Variations in discharge, sediment load and sediment composition in the River Nile and its main tributaries

Figure 1.1 shows variations in discharge, sediment load and sediment composition along the River Nile and its main tributaries.

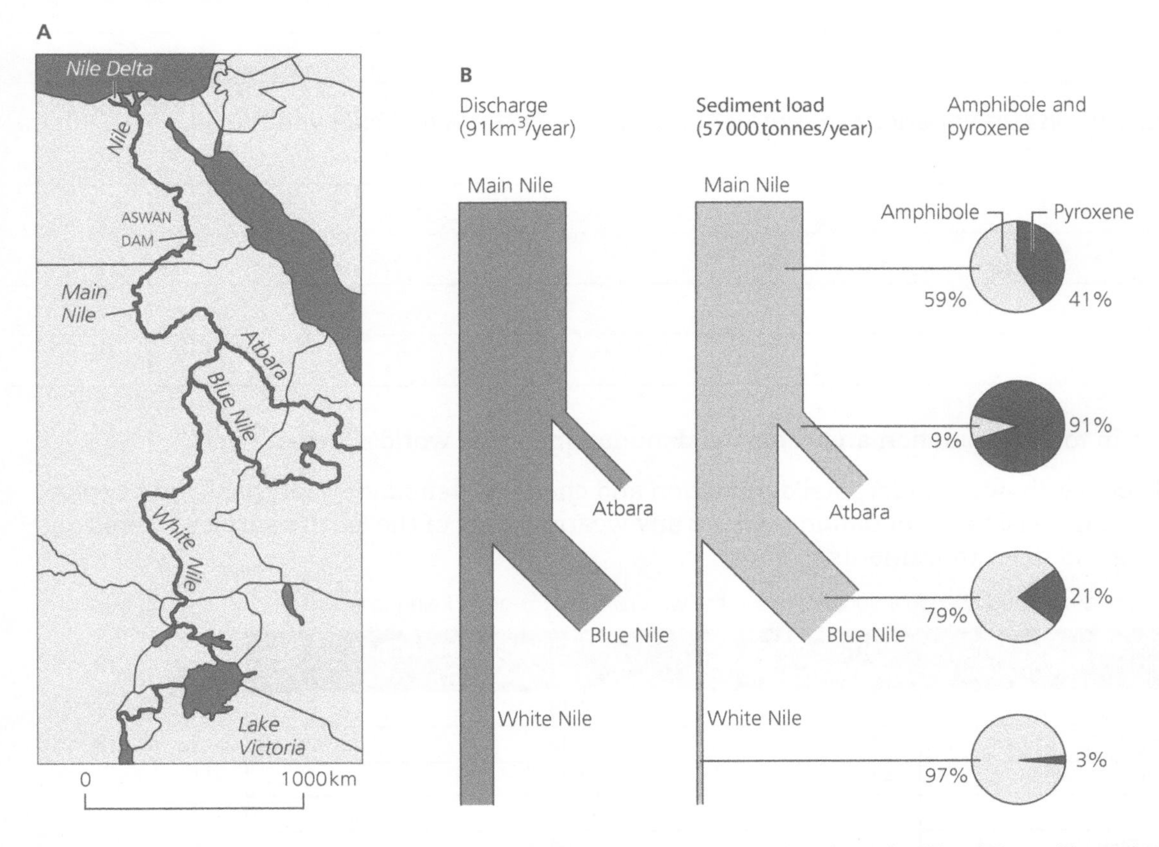

Figure 1.1 Variations in discharge, sediment load and sediment composition along the River Nile and its main tributaries

a Identify the three methods of visual representation used in Figure 1.1 and briefly describe one advantage and one disadvantage of each of the three methods.

Method	Advantage	Disadvantage

b Compare the relative contribution of each of the main tributaries to the flow of the Nile.

c Contrast the contribution to the sediment load with the discharge of each of the tributary rivers.

d Compare the composition of the load of the three tributaries.

e Suggest reasons for variations in sediment load between the White Nile and the Blue Nile/Atbara.

5 Variations in total denudation and chemical denudation for the world's largest rivers

The table below shows data for total denudation and chemical denudation for the world's twelve largest drainage basins. (Denudation refers to any wearing away of the Earth's surface by erosion, weathering and mass movements.)

Table 1.5 Denudation and chemical denudation for the world's twelve largest drainage basins

Drainage basin	Total denudation (mm/1000 years)	Chemical denudation (as % of total)
Amazon	70	18
Zaire (Congo)	7	42
Mississippi	44	20
Nile	15	10
Paraná (La Plata)	19	28
Yenisei	9	80
Ob	7	70
Lena	11	81
Yangtze	133	28
Amur	13	22
Mackenzie	30	33
Volga	20	64

a Suggest a definition for the term *chemical denudation.*

b Identify the rivers with the highest:

i rate of total denudation

ii contribution of chemical denudation.

c i Plot a scatter graph on the grid below to show the relationship between total denudation and contribution of chemical denudation.

ii Describe the main features of the graph that you have drawn.

____

____

____

____

iii Suggest reasons why some rivers have high rates of denudation.

____

____

____

____

Topic 2 Atmosphere and weather

1 The impact of human activities on global warming and cooling

Figure 2.1 shows how human activities and natural processes can have a warming or cooling effect on the climate.

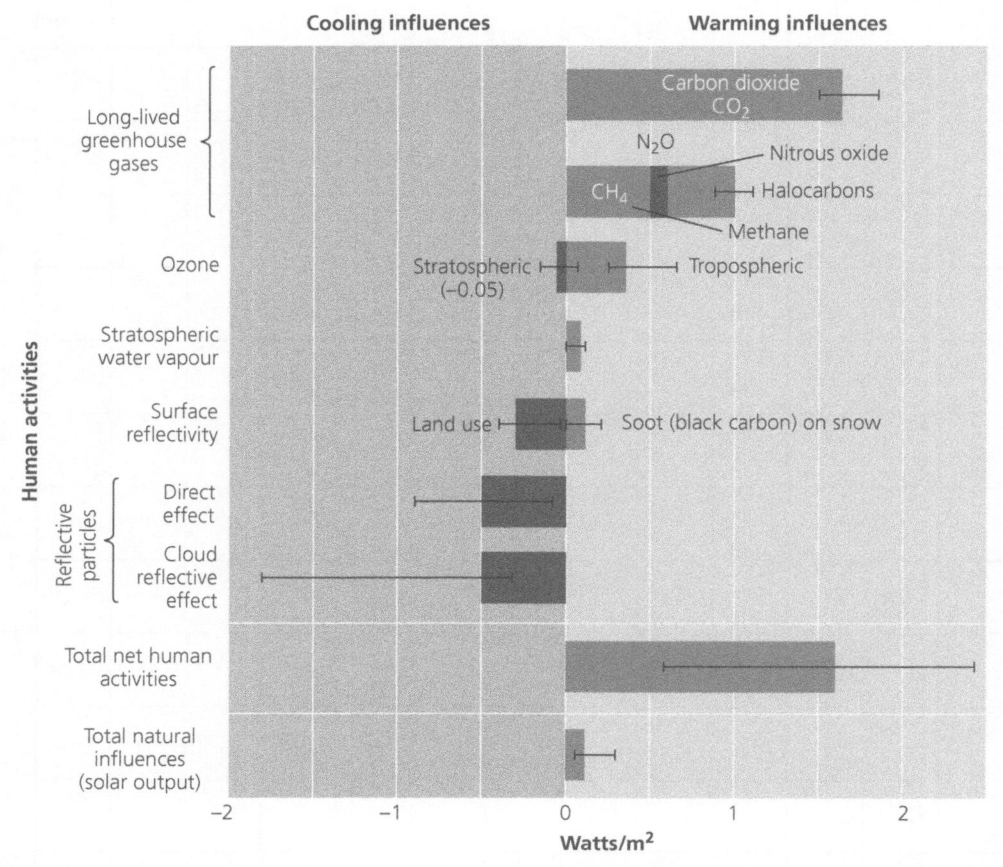

Figure 2.1 Box and whiskers diagram to show the impact of selected activities on warming (positive values) or cooling (negative values)

a i Identify the factor that had the greatest warming effect.

____

ii Identify the factor with the greatest cooling effect.

____

iii Identify two human activities that have both a warming and a cooling effect.

____

iv Which of these has the greatest warming effect?

____

v Compare the total natural influence with total net human influence.

____

____

b i Identify the factor with the greatest range of uncertainty.

____

ii Suggest why the range of uncertainty may be so high.

____

____

iii Identify one natural factor that may lead to cooling.

____

iv Suggest why this natural factor may have a limited impact on global cooling.

____

____

2 Variations in rainfall and relief in the Ganga–Brahmaputra–Meghna Basin

Figure 2.2 shows average monthly rainfall at selected stations in the Ganga–Brahmaputra–Meghna Basin.

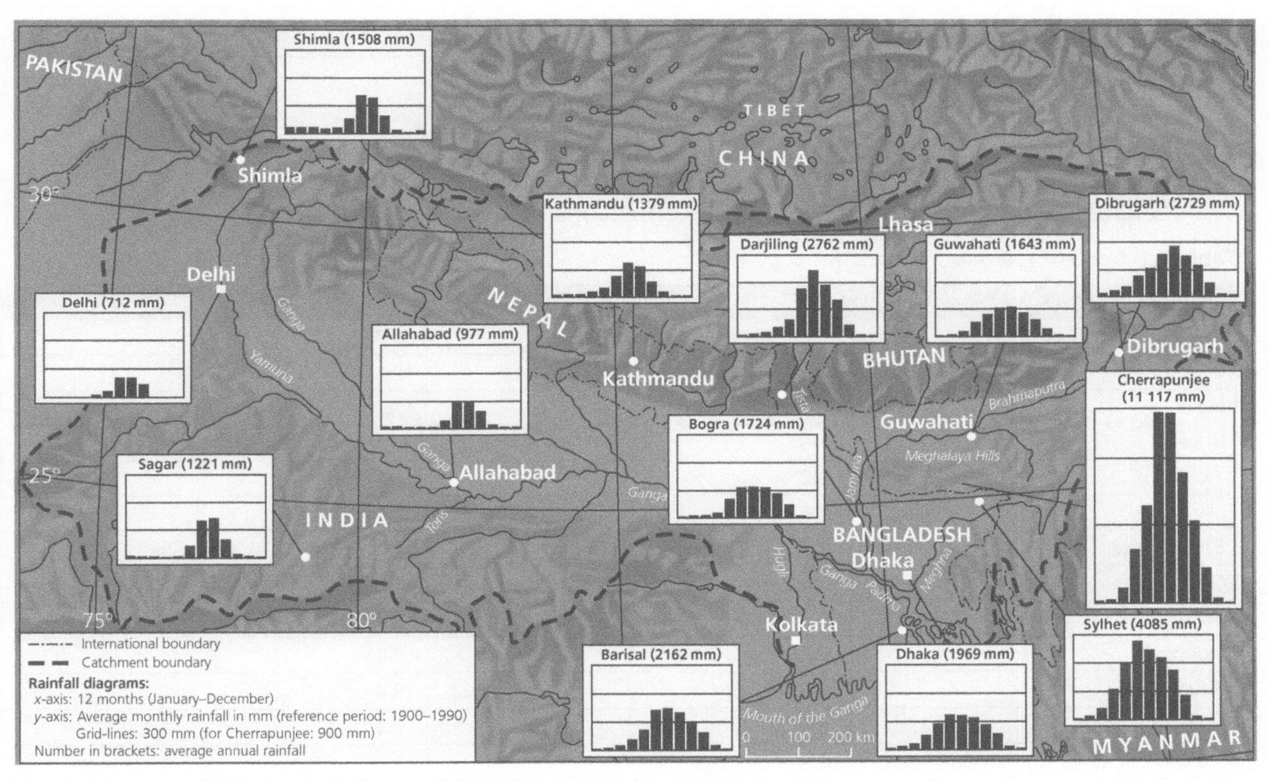

Figure 2.2 Average monthly rainfall in the Ganga–Brahmaputra–Meghna Basin

a Identify the techniques used to display the data.

____

____

b State one advantage and one disadvantage of each of the techniques shown on Figure 2.2.

Technique	Advantage	Disadvantage

c Describe the variations in rainfall, as shown in Figure 2.2.

d Suggest how this pattern may contribute to flooding.

3 CO_2 emissions for selected countries

The table below provides data for world CO_2 emissions in 2013.

Table 2.1 World CO_2 emissions in 2013

Country	Largest emitters, billion tonnes (rounded)	Emissions per person (tonnes, rounded)
China	10.2	6.6
USA	5.4	12.1
India	2.1	1.7
Russia	1.9	8.4
Japan	1.2	10.1
Germany	0.8	10.1
South Korea	0.7	10.1
Canada	0.6	11.2
Brazil	0.6	2.4
Indonesia	0.6	2.0
Saudi Arabia	0.5	12.4
United Kingdom	0.5	7.5
Mexico	0.4	3.9
Iran	0.3	5.2
Australia	0.2	12.5

a i Construct two bar charts to show the largest emitters and emissions per person by country.

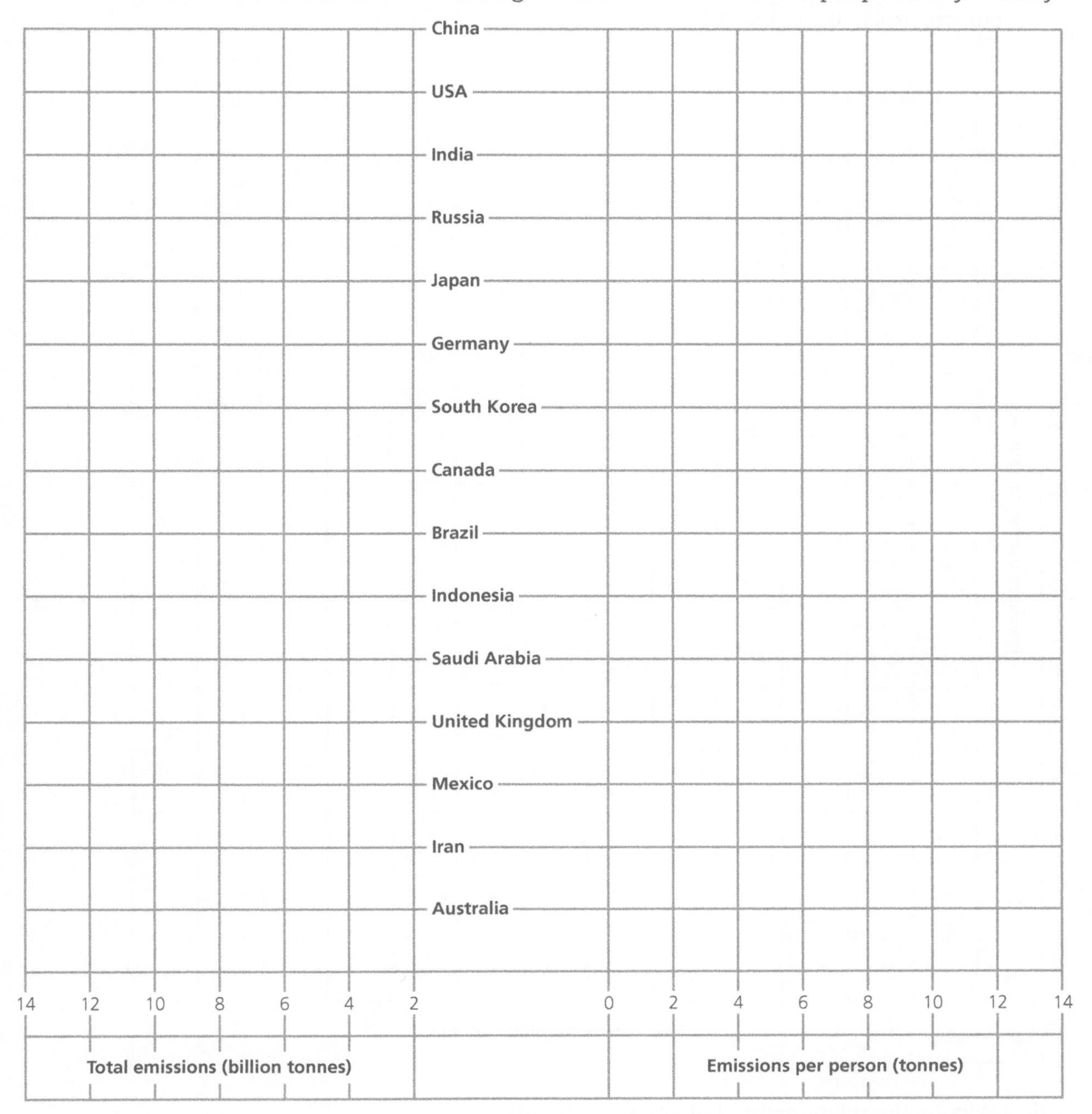

ii Identify the main characteristics of the countries with the top five emissions of carbon dioxide.

____

____

____

iii Identify the countries with the top four emissions per person.

____

iv State the likely characteristics of the countries with the top four emissions per person.

____

____

____

b i Draw a scatter graph to show the relationship between total emissions and emissions per person. Add a line of best fit.

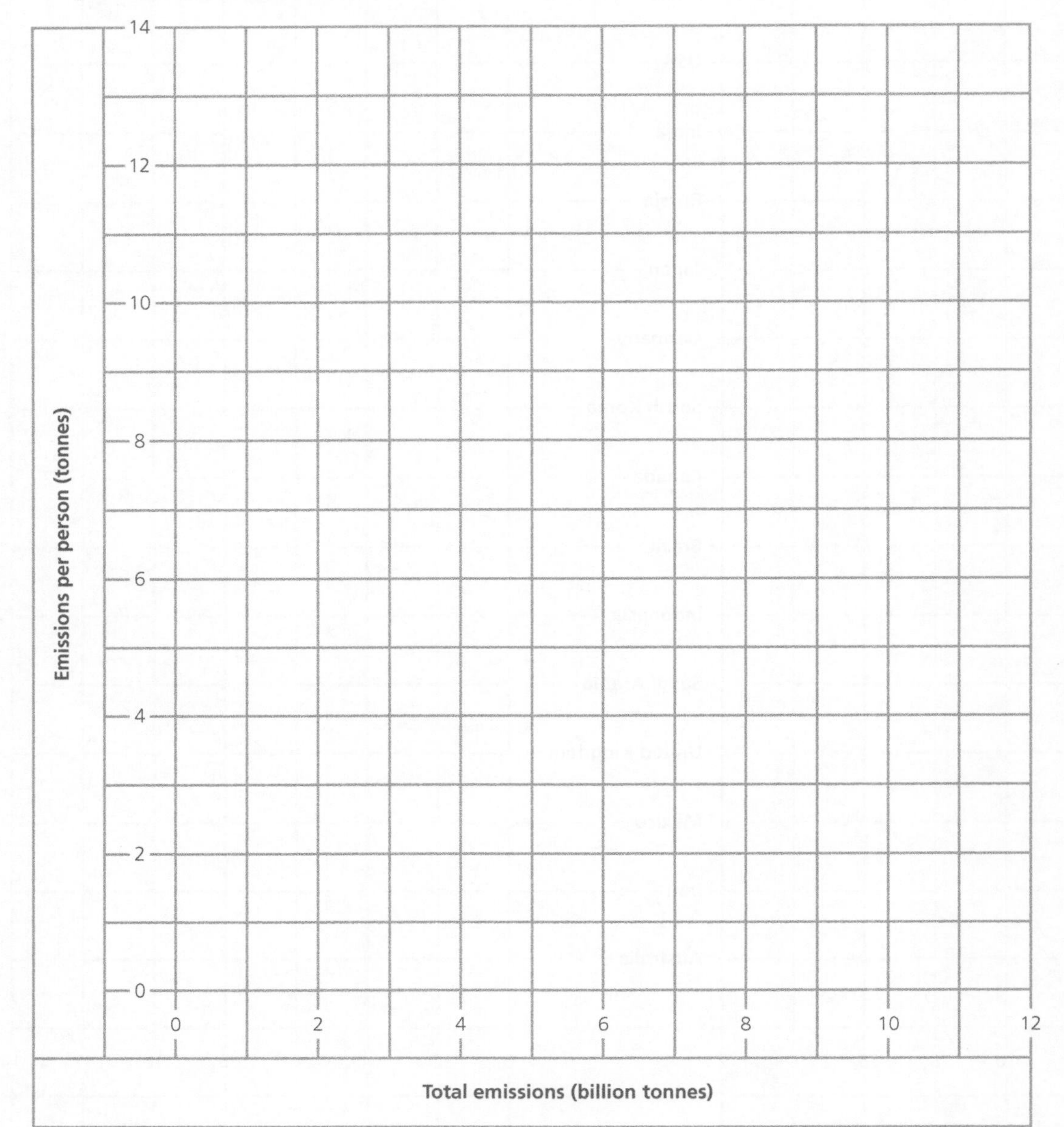

ii Describe the relationship that you have drawn.

iii Suggest reasons for the relationship.

iv Identify one strength and one weakness of the technique that you have used.

v State one way in which the technique you have used could be improved.

4 Energy budgets for Svalbard, Norway

Study Figure 2.7 on page 32 of the *Cambridge International A and AS Level Geography* textbook (second edition), which shows energy budgets for Svalbard, Norway.

a Identify the type of graph used to show the energy budgets.

b State one advantage of the graph.

c Identify one disadvantage of the method used.

d Explain the significance of the horizontal line on the graph.

e Identify the main features of the energy budget, as shown in the diagram.

Topic 3 Rocks and weathering

1 Tectonic plates and plate boundaries

Study Figure 3.4 on page 62 of the textbook, which shows plate boundaries.

a Identify the seven major tectonic plates.

b Identify three minor plates in North America/Caribbean.

c Describe the processes taking place along the Mid-Atlantic Ridge.

d Describe the processes taking place along the western side of South America.

e Draw an annotated diagram of a constructive plate boundary in the space below.

f Draw an annotated diagram of a destructive plate boundary in the space below.

g Study Figure 3.6a on page 64 of the textbook. Outline the evidence for sea-floor spreading.

2 Weathering and climate

Study Figure 3.13 on page 70 of the textbook, which shows Peltier's diagram of variations of types and rates of weathering with temperature and precipitation.

a State the conditions required for strong chemical weathering.

b Suggest two reasons why there is strong chemical weathering in the environment you have identified in answer to question **a**.

c State the conditions required for slight chemical weathering.

d Suggest reasons why there is some (slight) chemical weathering below 0 °C.

e Identify the type of weathering likely to occur when the mean annual temperature is 5 °C and the mean annual rainfall is 1000 millimetres.

f State the conditions required for strong mechanical weathering.

g Suggest why the rate of mechanical weathering decreases below –12 °C despite plentiful precipitation.

h Identify an alternative scale to mean annual temperature that might help explain the rate of mechanical weathering.

i Explain why this might not be an appropriate scale for all types of mechanical weathering.

3 Mass movements

Study Figures 3.17 and 3.18 on page 73 of the textbook, which show a classification of mass movement and speed of mass movement.

a Identify the type of graph used in Figure 3.17.

b State the type of scale used in Figure 3.17.

c Using Figure 3.17, describe the characteristics of slides and flows.

d Using Figure 3.17, describe solifluction.

e Using Figure 3.18, describe solifluction.

f Identify the type of scale used in Figure 3.18.

g State one advantage of the type of scale used in Figure 3.18.

h State the range of speed for earthflows/mudflows and for landslides.

i Using your answer to question **h**, comment on the accuracy of Figure 3.17.

j How could Figure 3.17 be made more accurate?

4 Plate boundaries and volcanic activity

The diagram below shows a model of volcanic activity at various plate boundaries.

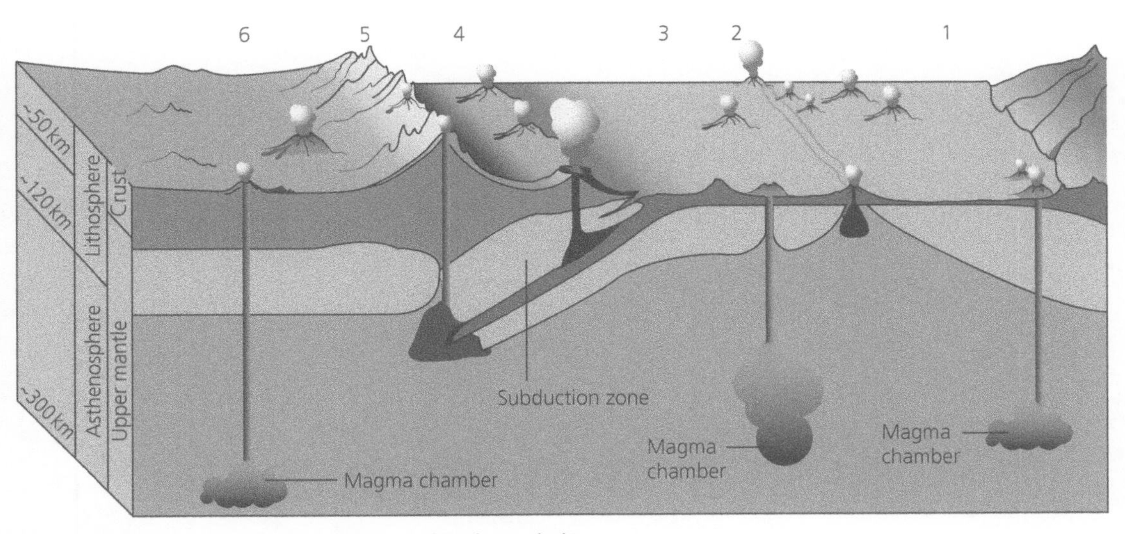

Figure 3.1 Model of volcanic activity at plate boundaries

a Choose labels for each of the sites 1–6. Write the appropriate number next to each of the options below (one option needs two numbers):

i continental hot spot ________

ii oceanic hot spot ________

iii island arc ________

iv fold mountains ________

v mid-ocean ridge ________

b Identify the tectonic plate boundaries at sites 2 and 5.

c Suggest how the magma may differ between sites 2 and 5.

d Using Table 3.1 on page 61 of the textbook, which shows a comparison of oceanic crust and continental crust, state how the crust at sites 5 and 6 may differ from that at sites 2 and 3 of Figure 3.1 above.

e Draw an annotated diagram to show a conservative plate boundary in the space below.

Paper 2 Core Human Geography

Topic 4 Population

1 Calculating basic demographic measures

Look at the data provided in the table below.

Table 4.1 Demographic data for six countries

Country	Birth rate	Death rate	Net migration rate (per 1000 population)	Rate of natural change (%)	Population change per year (%)
Canada	11	7	8		
USA	13	8	3		
Singapore	9	5	12		
Ireland	15	7	–7		
Spain	9	8	–4		
Greece	9	11	–4		

Source: Selected data from the 2014 World Population Data Sheet

a Explain how birth and death rates are calculated.

b Suggest why there are concerns about the accuracy of birth- and death-rate data for some countries.

c Calculate the rate of natural change (as a percentage) for each country and insert the figures in the appropriate column in the table.

d Define *net migration*.

e How is the rate of net migration expressed?

f Using your calculations of natural change and the figures for net migration, calculate the percentage population change per year for each country and insert the figures in the appropriate column.

g On the grid below, draw a median line bar graph to illustrate the net migration rate for the six countries. Use a scale of 1 centimetre for every 2/1000 net migration.

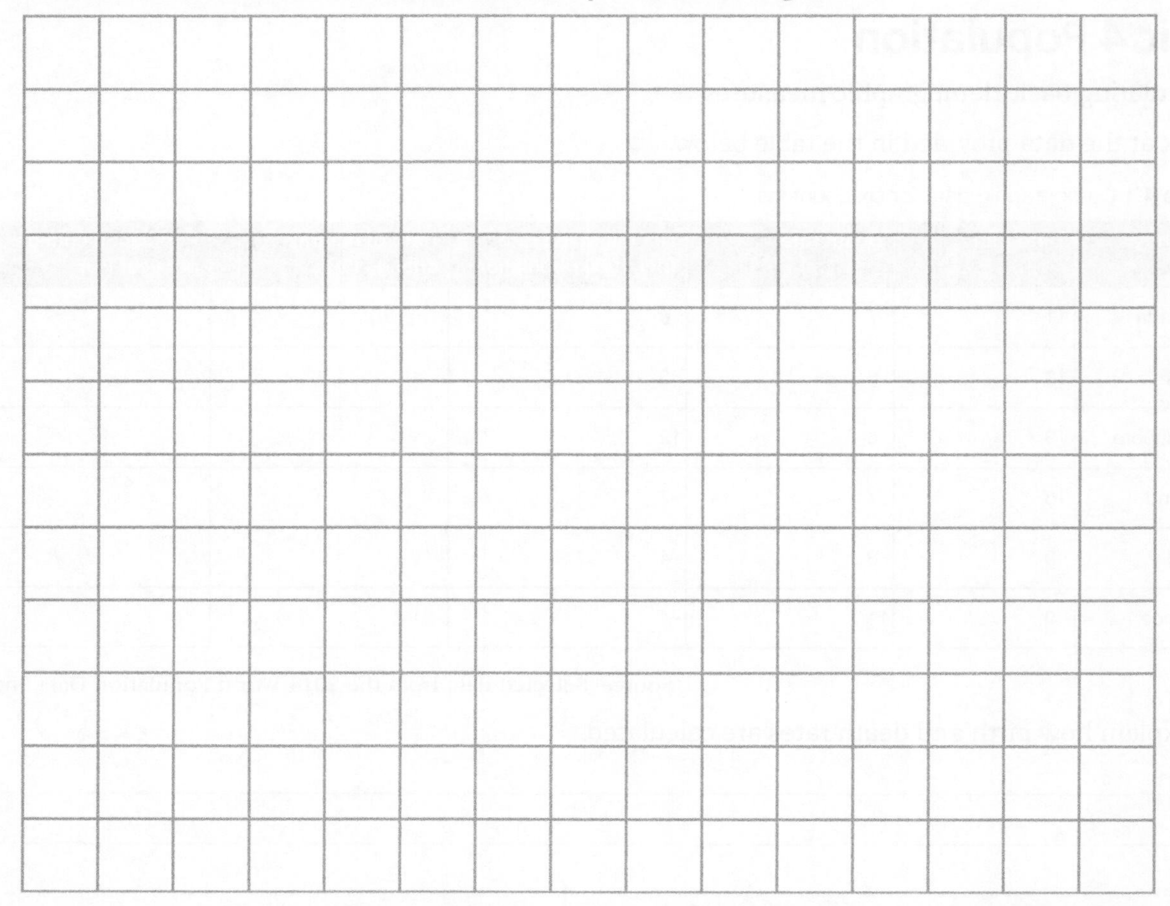

h If a country with a population of 60 million people records 720 000 births and 540 000 deaths in a year, calculate the birth rate and the death rate for that country.

____

____

2 Calculating and assessing the value of the dependency ratio

a What is the dependency ratio and how is it calculated?

____

____

____

____

____

____

b Suggest one reason why the dependency ratio is only a very broad measure of economic dependency with regard to population structure.

____

____

____

____

c Calculate the dependency ratio from the data provided for the countries listed in the table below and insert the figures in the appropriate column.

Table 4.2 Percentage populations <15 years and 65 and over

Country	% population <15	% population 65 and over	Dependency ratio
Niger	50	3	
Tanzania	45	3	
Bolivia	35	5	
Canada	16	15	
France	18	18	
Bulgaria	14	20	

d What does a dependency ratio of 70 mean?

e How can two countries have the same dependency ratio, but face very different demands on the resources at their disposal?

f How might a country attempt to reduce its:

i youth dependency ratio

ii elderly dependency ratio?

3 Geographical models and demographic transition

a What is a *geographical model?*

b Apart from the model of demographic transition (referenced in the following question), name two models you have studied in physical geography and two models you have studied in human geography.

c Draw a diagram of the model of demographic transition in the space below. Insert annotations referring to at least two characteristics for each stage of the model.

d What are the main criticisms of the standard model of demographic transition?

e State and briefly explain one alternative model of demographic transition.

f Compare the standard model of demographic transition with Figure 4.22 on page 97 of the *Cambridge International A and AS Level Geography* textbook (second edition), which shows the demographic history of England and Wales since 1700.

4 Analysing age/sex structure diagrams 1

Look at the four age/sex structure diagrams in Figure 4.14 on page 91 of the textbook (Niger, Bangladesh, UK, Japan).

a How many years of age are generally represented by each bar on an age/sex structure diagram?

____

b How does this change with a more detailed age/sex structure diagram?

____

c How is gender indicated on an age/sex structure diagram?

____

____

d Measure and state the population under 5 for each country.

i Niger: ______________

ii Bangladesh: ______________

iii UK: ______________

iv Japan: ______________

e In which stage of demographic transition is each of the four countries?

i Niger: __________

ii Bangladesh: __________

iii UK: __________

iv Japan: __________

f The x-axis (horizontal axis) of an age/sex structure diagram can show data in either absolute or relative form. Explain the difference between the two types of data.

____

____

g State the main advantage of using relative data when comparing the age/sex structure diagrams of different countries.

____

____

____

5 Analysing age/sex structure diagrams 2

a In the space below, draw age/sex structure diagrams to represent the typical population structures of HICs and LICs.

b Divide each diagram into three age groups in terms of the dependency ratio, clearly labelling each group.

c Describe the differences between the two age/sex structure diagrams for each age group.

d In the space below, draw an annotated age/sex structure diagram of a rural area in a LIC that has experienced a high rate of out-migration.

e How would you recognise a high rate of in-migration on an age/sex structure diagram of a large urban area?

6 Assessing the relationship between two variables

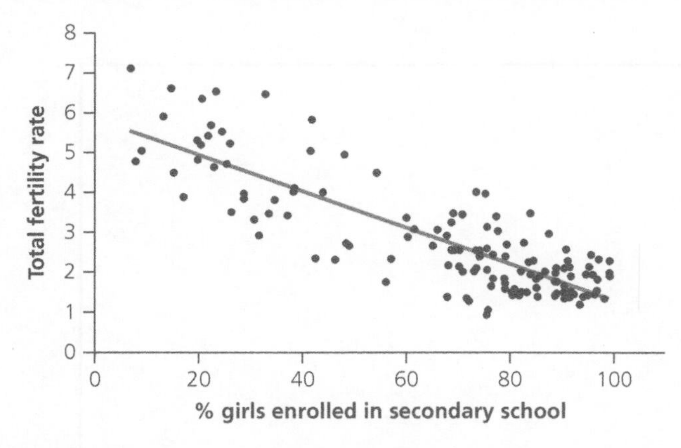

Figure 4.1 Comparison between female secondary education and total fertility rates

a If you were presenting an analysis of the relationship between two variables, for example the infant mortality rate and GDP per person, you might use statistical analysis, qualitative analysis and graphical analysis. Briefly state the meaning of each form of analysis and the logical order of presentation.

b Look at Figure 4.1 above. What type of graph is this?

c Name the straight line drawn on the graph.

d What is the objective of this straight line?

e Describe the relationship between the two variables shown by the graph.

f What is the name given to plots on the graph that are located far from the general trend of the graph?

g In the space below, draw simple diagrams to show the difference between a perfect positive correlation and a perfect negative correlation.

h Name a statistical technique that could be used to determine the degree of relationship between two sets of data.

---

i What can the statistical technique you have named add to the data illustrated by a scatter graph?

---

7 Demographic contrasts illustrated by divided bar graphs

Figure 4.11 on page 90 of the textbook is a divided bar graph showing contrasts in the causes of death between HICs and LICs.

a Explain the construction of the divided bar graphs shown in Figure 4.11.

b How effective is this technique in showing the contrast in the causes of death between HICs and LICs?

c How different would the two divided bars in Figure 4.11 look if absolute data rather than relative data had been used?

d Compare the causes of death for HICs and LICs shown in Figure 4.11.

e Describe another type of bar graph that could be used to illustrate this data.

Topic 5 Migration

1 The age distribution of international migrants in LICs/MICs and HICs

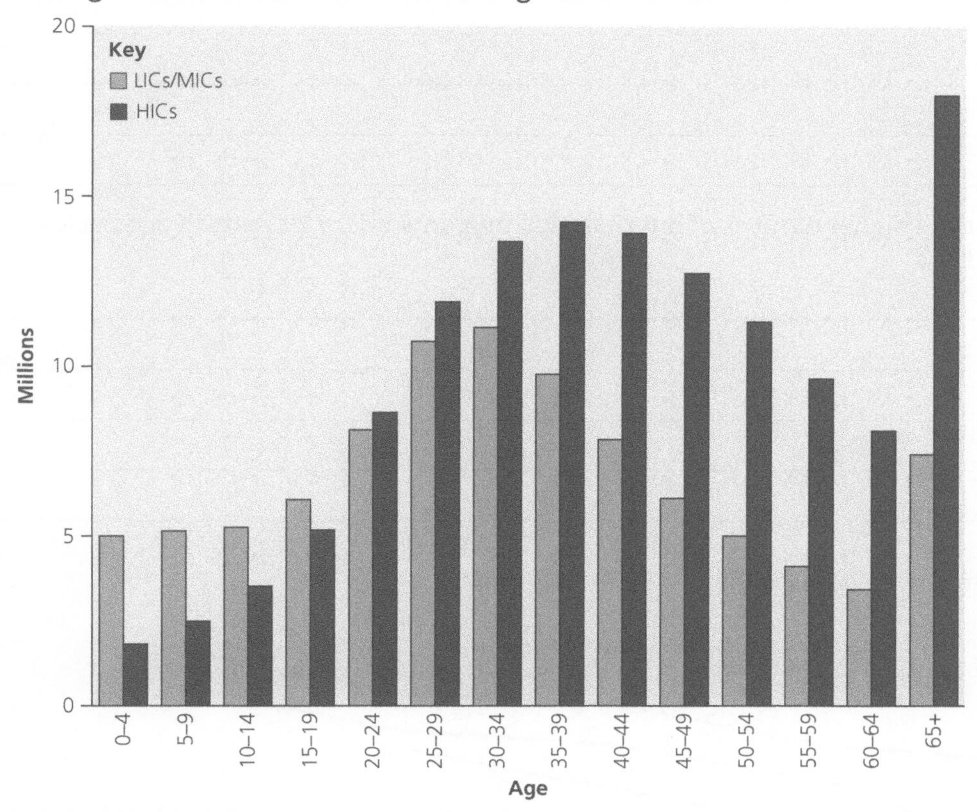

Figure 5.1 Age of international migrants in LICs/MICs and HICs

In 2013, three-quarters of all international migrants were between the ages of 20 and 64. Of the 171 million international migrants in this age group, 61 per cent resided in HICs. This distribution has changed little since the year 2000.

a Explain the organisation of the horizontal and vertical axes on the graph.

____

____

____

____

b Describe the way the data has been illustrated.

____

____

____

____

c Suggest an alternative way of organising the vertical scale.

____

____

d Compare the age distribution of international migrants in LICs/MICs and HICs.

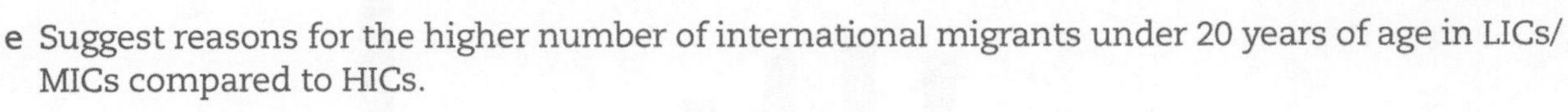

e Suggest reasons for the higher number of international migrants under 20 years of age in LICs/MICs compared to HICs.

2 Using time-series line graphs to illustrate immigrant earnings

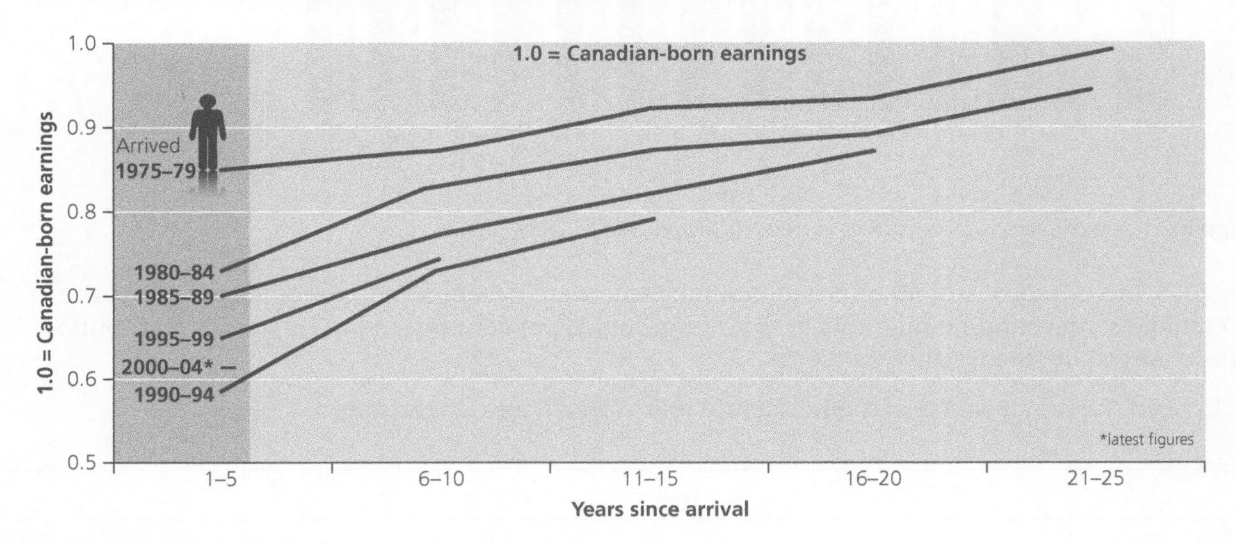

Figure 5.2 Male immigrants' earnings over time after arrival in Canada

a Describe the arrangement of the vertical (y) scale on the graph.

b How is the horizontal (x) scale organised?

c What does the different length of the lines on the graph indicate?

d Describe the information on male immigrants' earnings in Canada illustrated by the graph.

e Suggest possible reasons for the different levels of earnings illustrated.

Topic 6 Settlement dynamics

1 Rural settlement: location and change

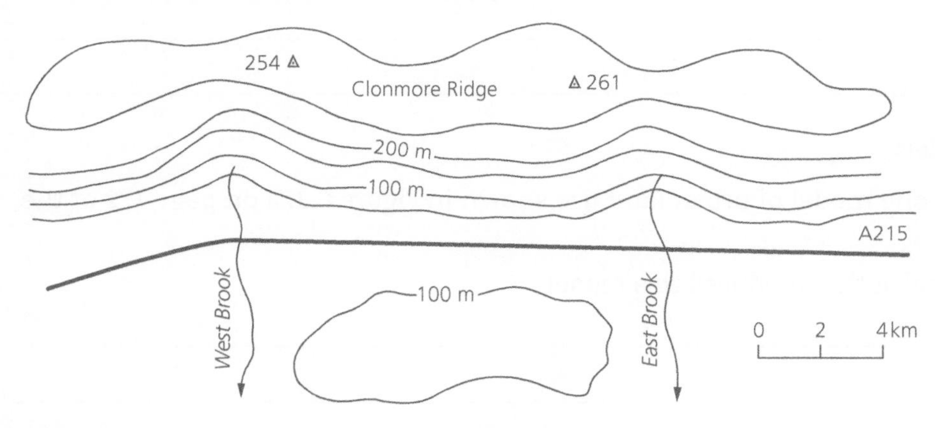

Figure 6.1 Contour-style map for locating rural settlements

a On the diagram, use the letters A and B to mark the likely sites of two villages originally located in the medieval period.

b Justify the sites you have chosen.

c How might these rural settlements expand with modern population increase? Show what this could look like on the diagram.

d Why are some rural areas affected by rural depopulation?

e Why are other rural areas expanding rapidly in terms of population?

2 Investigating urban models

Look at the concentric zone model of urban land use shown in Figure 6.32a on page 172 of the textbook.

a Suggest why the original author produced this model.

b Describe the pattern of land use shown by the model.

c What were the main assumptions upon which the model was based?

d Explain one advantage and one disadvantage of this model.

e Suggest how this model could be improved.

3 The use of semi-log and double-log (log–log) graphs in urban analysis

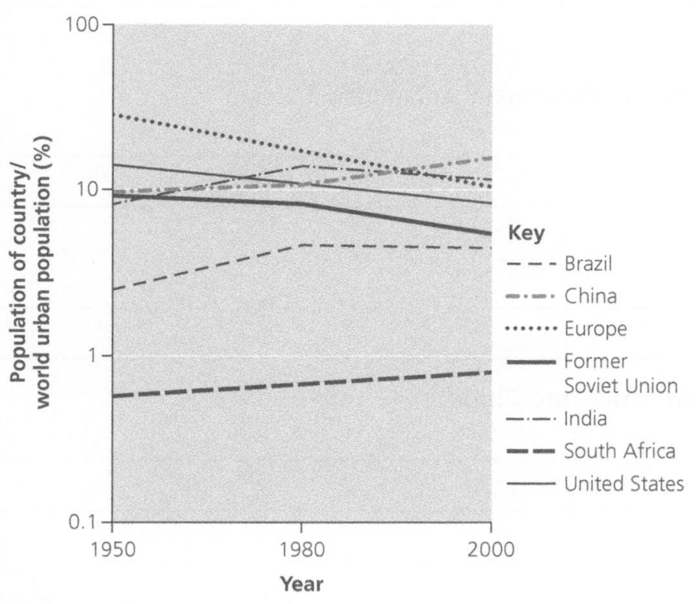

Figure 6.2 Semi-log graph: population of country/region as a percentage of world urban population

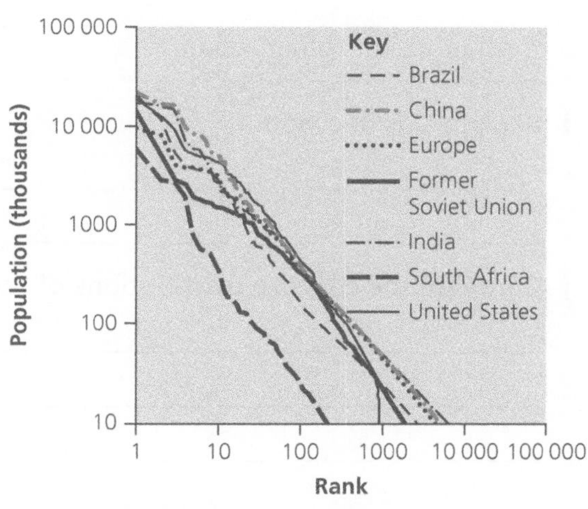

Figure 6.3 Double log (log–log) graph: city size distributions in seven countries around 2010

Look at Figure 6.2.

a What type of scale has been used on the vertical (y) axis?

b Describe and explain the applicability of this type of scale.

c What type of scale has been used on the horizontal axis?

d How could you question the accuracy of the way this scale has been used?

e The graph has been constructed using data for how many years?

f State the years in question.

g Which countries/regions have recorded a decline in their share of the world's urban population between 1950 and 2000?

Look at Figure 6.3.

h How do the scales used on this graph differ from those used in Figure 6.2?

i Explain this decision.

j Compare the city size distributions of South Africa and China.

Paper 3 Advanced Physical Geography Options

Topic 7 Tropical environments

1 Tropical rainforest vegetation and altitude

The table below shows the number of Dipterocarp species (a type of tree) found at different altitudes in Brunei.

Table 7.1 The number of Dipterocarp species found at different altitudes in Brunei

Altitude (m)	Number of species
0–99	134
100–199	122
200–299	120
300–399	110
400–499	71
500–599	67
600–699	43
700–799	34
800–899	28
900–999	26
1000–1099	21
1100–1199	17
1200–1299	13
1300–1399	11
1400–1499	2
1500–1599	1

a Choose an appropriate technique to show the relationship between the number of dipterocarp species and altitude and draw this in the grid below.

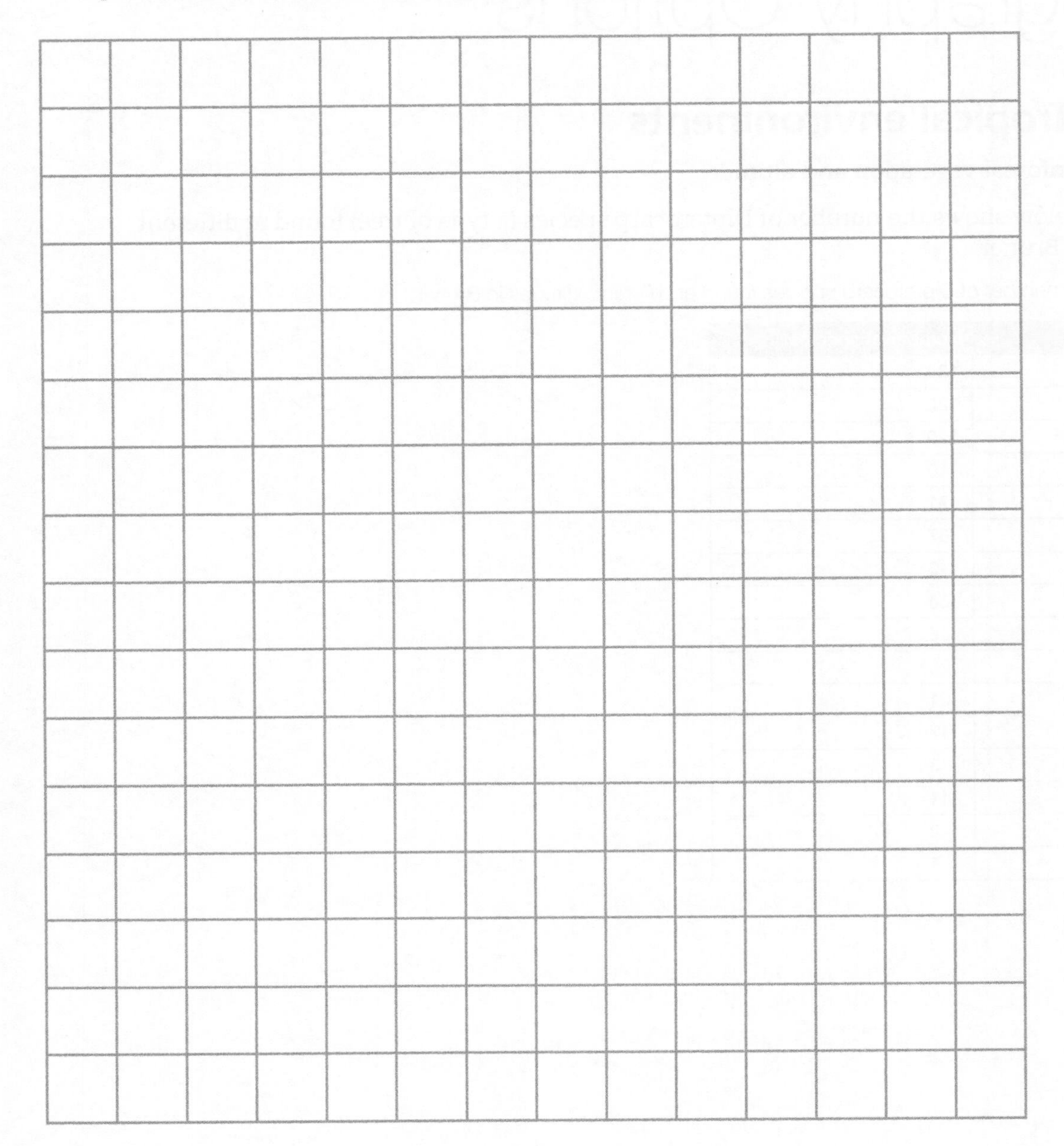

b State one advantage of the method that you have used.

c Describe the relationship that you have drawn.

d Suggest reasons for the relationship that you have described.

2 Natural and disturbed nutrient cycles in the tropical rainforest

Study Figures 7.37 and 7.38 on page 212 of the *Cambridge International A and AS Level Geography* textbook (second edition), which show generalised nutrient cycles and nutrient cycles for a tropical rainforest under natural conditions and following farming.

a State the techniques used to show nutrient cycles.

b Identify the largest store of nutrients in the tropical rainforest.

c State the store with the least amount of nutrients in the tropical rainforest.

d Identify the two smallest flows of nutrients in the tropical rainforest.

e Describe the changes in the nutrient cycle of tropical rainforests following the conversion of land use to farming.

f Compare the nutrient cycle in the savanna with that of the tropical rainforest.

3 Ecosystem services

The following table shows the assessment of the trend in the global state of ecosystem services.

Table 7.2 The trend in global ecosystem services

Service	Sub-category	Status	Notes
Provisioning services			
Food	Crops	↑	Substantial production increase
	Livestock	↑	Substantial production increase
	Capture fisheries	↓	Declining production due to over-harvest
	Aquaculture	↑	Substantial production increase
	Wild foods	↓	Declining production
Fibre	Timber	+/–	Forest loss in some regions, growth in others
	Cotton, hemp, silk	+/–	Declining production of some fibres, growth in others
Genetic resources		↓	Loss through extinction and crop genetic resource loss
Biochemicals, natural medicines, pharmaceuticals		↓	Loss through extinction, overharvest
Freshwater		↓	Unsustainable use for drinking, industry and irrigation; amount of hydro-energy unchanged, but dams increase ability to use that energy
Regulating services			
Air-quality regulation		↓	Decline in ability of atmosphere to cleanse itself
Climate regulation	Global	↑	Globally ecosystems have been a net sink for carbon since mid-twentieth century
	Regional and local	↓	Preponderance of negative impacts (e.g. changes in land cover can affect local temperature and precipitation)
Water regulation		+/–	Varies depending on ecosystem change and location
Erosion regulation		↓	Increased soil degradation
Water purification and waste treatment		↓	Declining water quality
Disease regulation		+/–	Varies depending on ecosystem change
Pest regulation		↓	Natural control degraded through pesticide use
Pollination		↓	Apparent global decline in abundance of pollinators
Natural hazard regulation		↓	Loss of natural buffers (wetlands, mangroves)
Cultural services			
Spiritual and religious values		↓	Rapid decline in sacred groves and species
Aesthetic value		↓	Decline in quantity and quality of natural lands
Recreation and ecotourism		↓	More areas accessible but many degraded

Source: World Bank World Development Report 2010, *Development and climate change*

a State whether the data provided in the table is qualitative or quantitative.

b Define the terms *qualitative* and *quantitative*.

c Define the term *ecosystem services*.

d Describe the general trend in ecosystem services.

e Identify the ecosystem services that are improving.

f Identify the ecosystem services that have a degree of uncertainty about their trend.

4 Tropical rainforests

The table below shows the mean maximum temperature and daily temperature range at two heights above the ground in a tropical rainforest.

Table 7.3 Mean maximum temperature and daily temperature range in a tropical rainforest

	Dry season		Wet season	
Sampling height (m)	0.7	24.0	0.7	24.0
Mean maximum temperature (°C)	29.7	33.9	26.8	30.9
Daily range	5.8	9.9	5.5	9.2

a Identify the height that has the highest temperature.

b Briefly explain why this height has the highest temperature.

c Identify the season with the highest temperatures.

d Suggest why this season has the higher temperatures.

e Identify the height with the highest daily range of temperatures.

f Suggest a reason for the daily range of temperatures being higher there than at the other site.

g Identify one factor about the data provided that is unusual for a rainforest location.

5 Rainforest vegetation

This photograph shows an area of rainforest in Borneo.

Figure 7.1 Rainforest in Borneo

a Describe the vegetation as shown in the photograph.

b Outline the adaptations of vegetation in the tropical rainforest.

c Identify three different vegetation layers in the tropical rainforest.

d Suggest how environmental conditions at the top of the forest may differ from those near the forest floor.

Topic 8 Coastal environments

1 Comparing sediment sizes on a beach and a sand dune

The following table shows the distribution of sediment sizes on a beach and on a sand dune.

Table 8.1 Distribution of sediment sizes on a beach and sand dune

Category	Diameter (mm)	Sediment name	Beach (%)	Sand dune (%)
1	1.01–2.00	Very coarse sand	3	–
2	0.51–1.00	Coarse sand	10	4
3	0.26–0.5	Medium sand	69	6
4	0.126–0.25	Fine sand	7	70
5	0.0625–0.125	Very fine sand	1	18
6	<0.0625	Silt	–	2

a Construct two histograms to show the distribution of sediment size on the beach and on the sand dune.

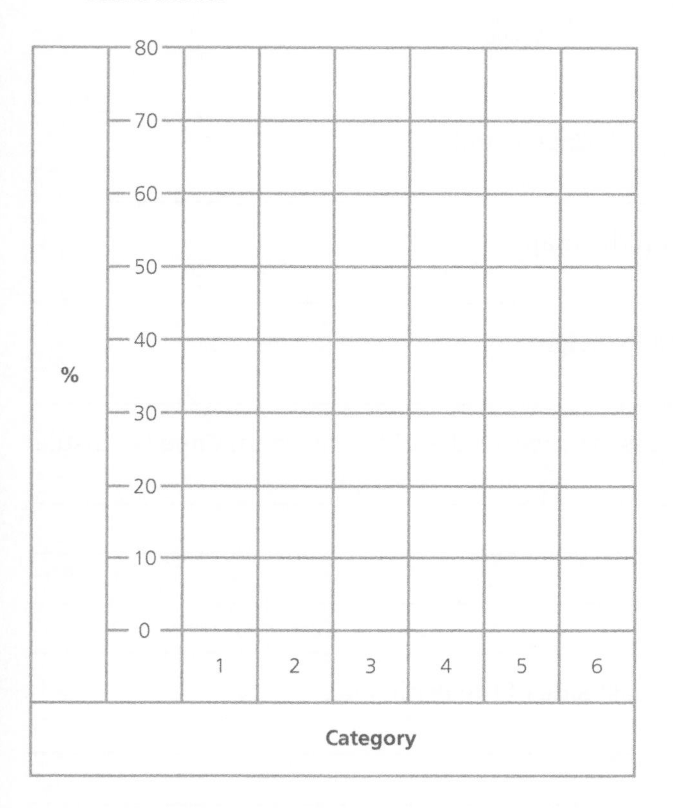

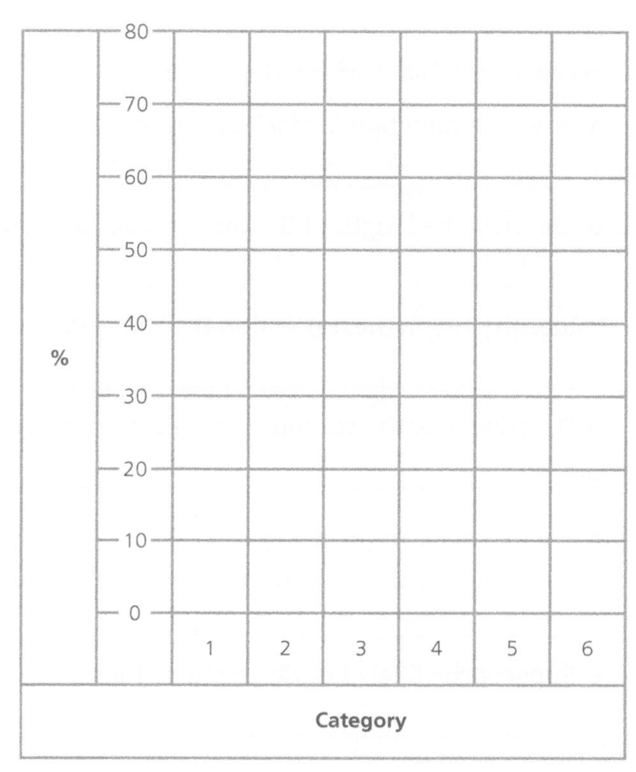

b Describe the main characteristics of the sediment-size distribution on the beach and on the sand dune.

c Suggest a reason for the differences you have shown.

2 Coastal landforms

The map and photograph (Figure 8.2) show a coastal landscape in South Africa.

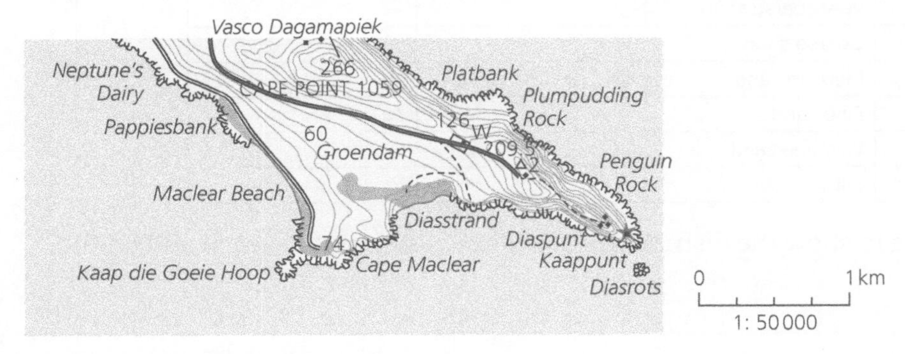

Figure 8.1 The Cape Peninsula, South Africa

a In which direction is Maclear Beach from Kaappunt (Cape Point)?

b Measure the length of the longest road shown on the map.

c Identify the highest point on the map, and state its height.

d Describe how the contour patterns differ on the east and west sides of the southern Cape Peninsula.

e Suggest the likely landforms to be found on the east side of the Peninsula.

f Study the photograph below and decide whether it shows Kaap die Goeie Hoop or Kaappunt.

Figure 8.2

g Explain how you made your choice.

h Draw an annotated sketch in the space below to show the main features of the landscape shown in the photograph.

3 Sea-level change around the USA

Study Figure 8.52 on page 257 of the textbook, which shows relative sea-level change in the USA.

a Identify the type of graph used.

b State one advantage of the method used.

c State one disadvantage of the method used.

d How does the use of colour aid the graph?

e Identify an alternative method to show the data.

f Using an atlas or the internet, locate the places mentioned in Figure 8.52. Comment on the geographic variation in sea-level changes as shown in the diagram.

4 Coral reefs

Study Figure 8.44 on page 249 of the textbook, which shows the world distribution of coral reefs.

a Describe the world distribution of coral reefs.

b Briefly describe one factor to help explain the distribution of coral reefs.

c Look at Figure 2.14 on page 37 of the textbook, which shows surface winds and pressure. Suggest one reason why there is a lack of coral reefs off the west coast of South America and the west coast of southern Africa.

d Suggest one reason why coral reefs are located relatively close to land masses.

e Suggest one reason why there is a lack of coral off the coast of South America near the equator, while there is coral off the north-east coast of Africa near the Tropic of Cancer.

Topic 9 Hazardous environments

1 Post-disaster recovery

The graphs below show stages of recovery in Antigua and Montserrat after the impact of Hurricane Hugo.

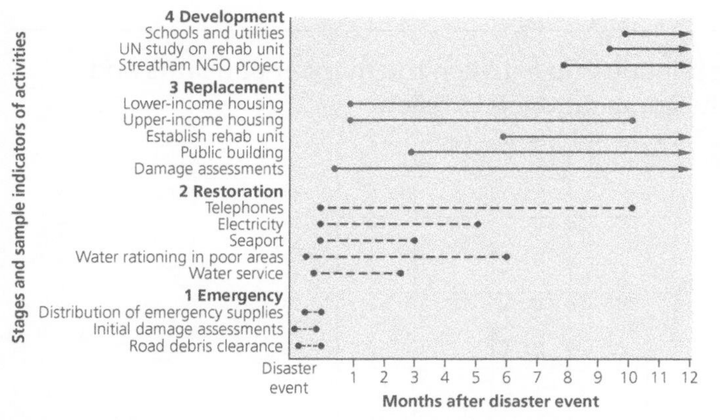

Figure 9.1 Recovery in Montserrat following Hurricane Hugo

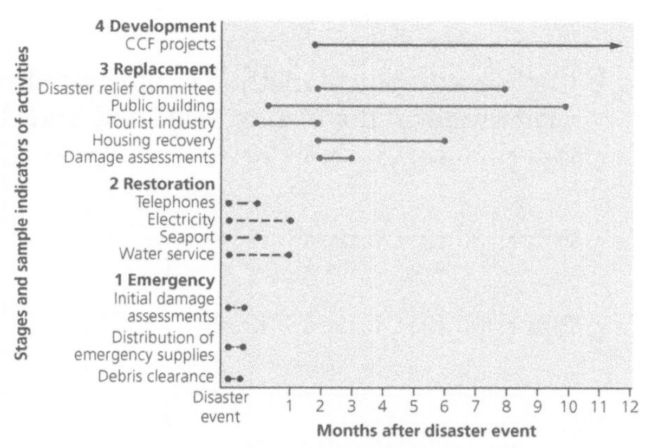

Figure 9.2 Recovery in Antigua following Hurricane Hugo

a Compare the length of the emergency aid in Antigua and in Montserrat.

____

____

b Compare the length of the restoration aid in Antigua and Montserrat.

____

____

____

____

c Identify possible reasons why the replacement phase in Montserrat took longer than in Antigua.

____

____

____

____

d Suggest reasons why the development phase required longer than twelve weeks.

____

____

____

____

2 Earthquakes – the relationship between magnitude and loss of life

Study Table 9.4 on page 268 of the textbook, which lists the world's worst earthquakes by death toll in the twenty-first century.

a Describe the relationship between the death toll and the Richter Scale.

b Choose a graphical method to illustrate the relationship between the number of deaths and magnitude on the Richter Scale and draw the graph on the grid below.

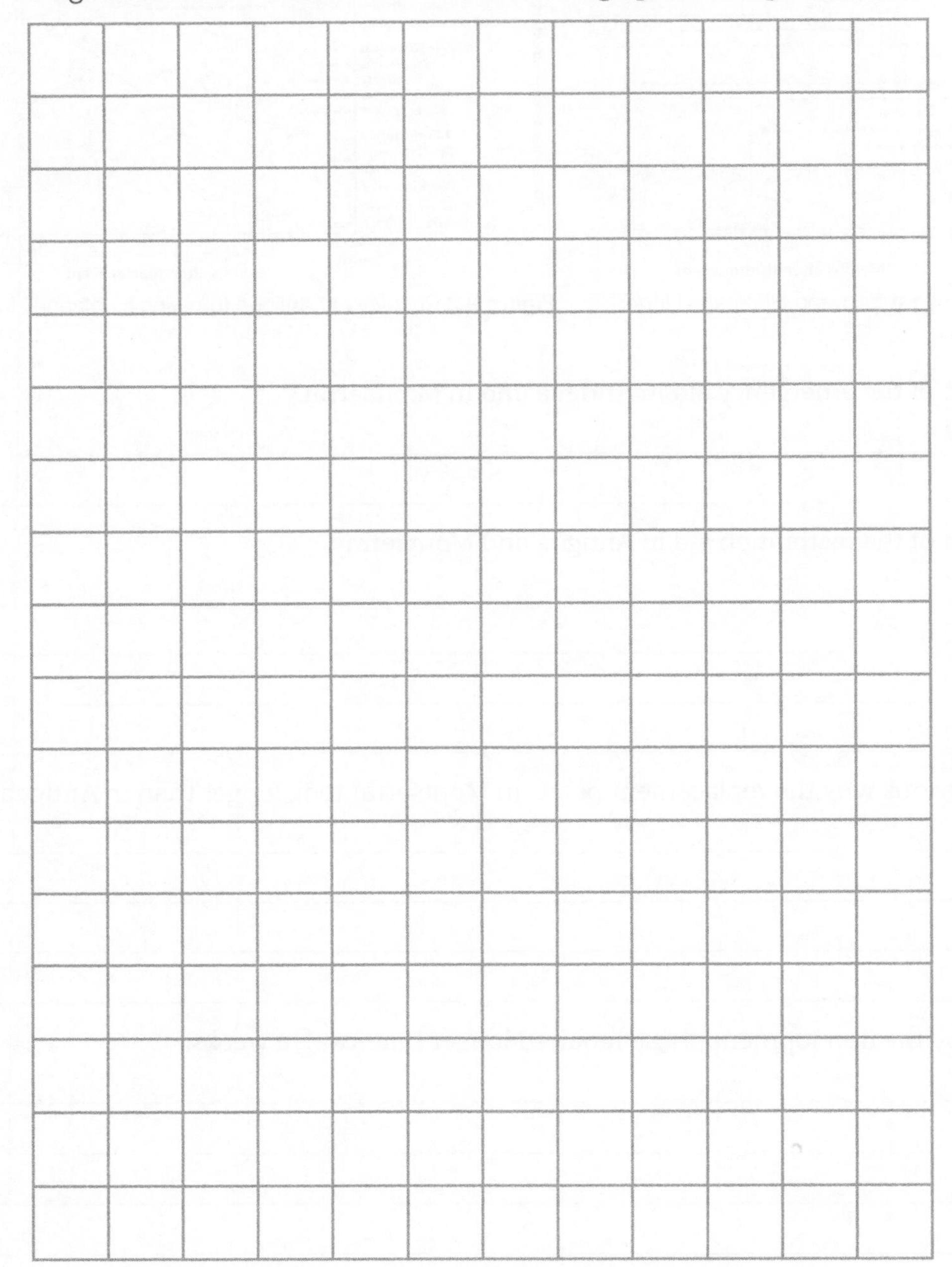

c Briefly explain two problems with using the data in Table 9.4.

d State one other investigation that could be studied using the data in Table 9.4.

e Study Figure 9.5 on page 271 of the textbook, which shows the increasing earthquake frequency associated with underground liquid-waste disposal in Rocky Mountain Arsenal, Colorado, USA. Identify the type of graph used.

f State when the maximum fluid injection occurred.

g Identify the month and year when most earthquakes occurred.

h Approximately how many earthquakes per month occurred during the period of no fluid injection?

i Identify two periods when there was a high fluid injection and high frequency of earthquakes.

3 Tropical storms

Study Figure 9.36 on page 293 of the textbook, which shows the distribution of tropical storms.

a Identify the **two** geographical techniques used on the map.

b Describe the distribution of tropical storms as shown on the map.

c Describe the paths taken by tropical storms in the western Pacific.

d Compare this with the paths taken by tropical storms in the southern Indian Ocean.

e Study Figure 9.41 on page 297 of the textbook, which shows the distribution of tornadoes in the USA. Describe the distribution and frequency of tornadoes as shown on the map.

f Identify three contrasts between the distribution and location of tropical storms and tornadoes.

4 Natural hazards, 2004–13

The table below shows the number of deaths attributed to different natural hazards between 2004 and 2013.

Table 9.1 Number of deaths attributed to natural hazards, 2004–13 (n/a indicates that no data are available)

	2004	2005	2006	2007	2008	2009	2010	2011	2012	2013	Total
Earthquakes and tsunamis	227 290	76 240	6 690	780	87 920	1 890	226 740	20 950	710	1 120	
Floods	7 370	5 750	5 850	8 570	4 030	3 530	8 570	6 140	3 580	9 820	
Mass movements	610	650	1 650	270	620	690	3 400	310	520	280	
Volcanic eruptions	2	3	5	10	20	n/a	320	3	n/a	n/a	
Hurricanes/ tornadoes	6 610	5 290	4 330	6 040	140 990	3 290	1 500	3 100	3 100	9 210	
Total natural disasters (including others)	**242 830**	**88 890**	**23 850**	**16 860**	**235 270**	**10 860**	**297 730**	**31 320**	**9 540**	**22 450**	

a Work out the total number of deaths for each hazard over the ten-year period and write them in the table.

b Work out the average number of deaths for each hazard per year.

c Identify the natural hazard that caused the greatest number of deaths.

d State the name of the natural hazard that caused the least loss of life.

e Compare the number of deaths by main cause of death in 2004 and 2012.

f Choose an appropriate technique to illustrate the data needed to answer question **e**.

g Describe the trend in the number of deaths due to natural hazards over time.

5 Worst volcanic hazards

The table below shows the results of a student's research into the world's top twelve most deadly volcanic eruptions.

Table 9.2 Top twelve most deadly volcanic eruptions

Rank	Volcano	Location	Year of eruption	Death toll	Major cause of death
1	Tambora	Indonesia	1815	92000	Ash fall, starvation
2	Krakatoa	Indonesia	1883	36417	Ash fall, tsunami
3	Mount Pelée	Martinique	1902	29025	Pyroclastic flows
4	Nevado del Ruiz	Colombia	1985	25000	Lahars
5	Unzen	Japan	1792	14300	Volcano collapse, tsunami
6	Laki	Iceland	1783	9350	Lahars
7	Kelut	Indonesia	1919	5110	Lahars
8	Galunggung	Indonesia	1882	4011	Lahars
9	Vesuvius	Italy	1631	3500	Lava flows, lahars
10	Vesuvius	Italy	79	3360	Ash falls, pyroclastic flows
11	Pandayan	Indonesia	1772	2957	Pyroclastic flows
12	Lamington	Papua New Guinea	1951	2942	Pyroclastic flows

Armstrong, D. *et al.*, 2008, *Geology,* Heinemann

a On the world map below, locate the twelve volcanoes in the table above. Choose a symbol/scale to indicate the number of deaths caused.

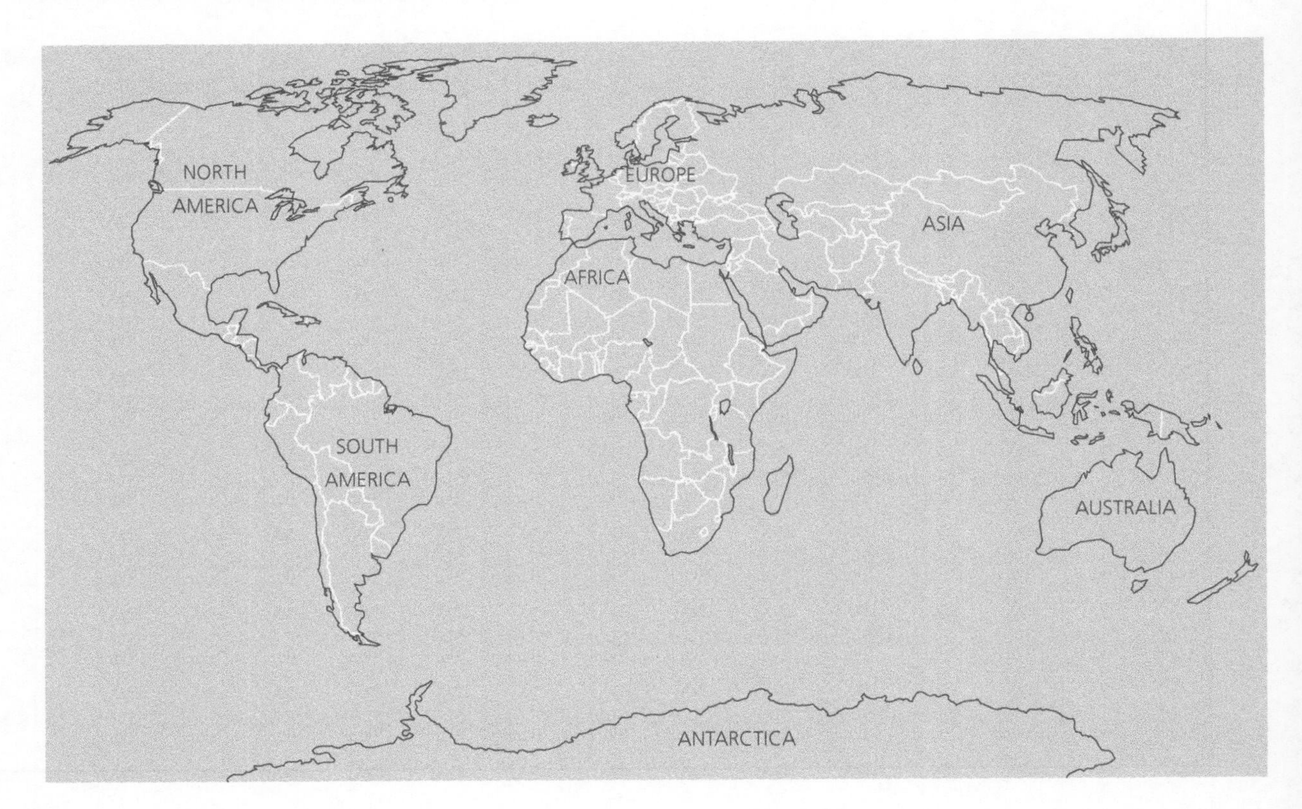

b Describe the distribution of volcanoes as shown on the map you have drawn.

____

____

____

____

c Draw a time line to show the date of the eruptions in Table 9.2 and the number of deaths caused.

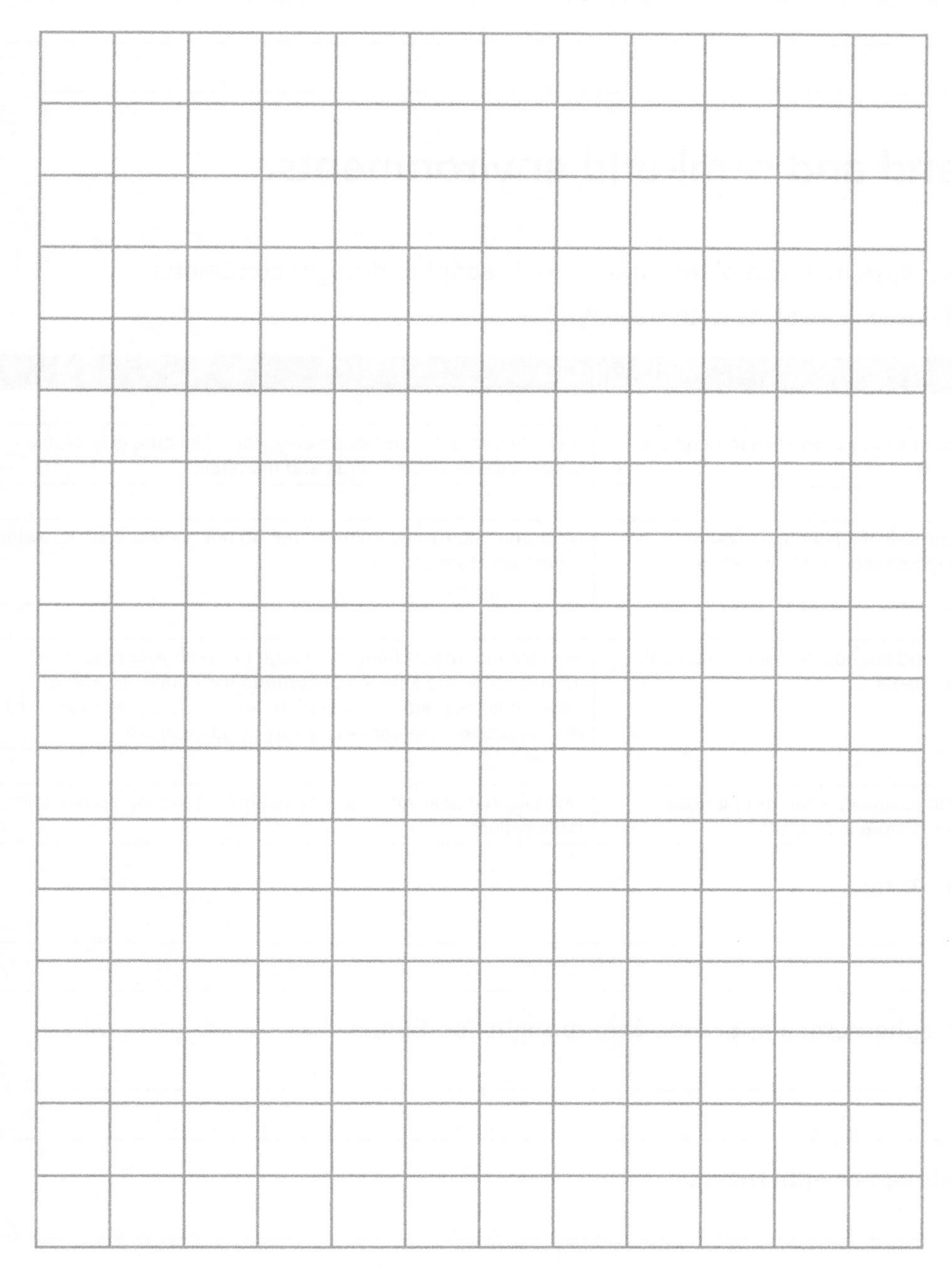

d Describe the variations in the frequency of volcanic eruptions and the number of deaths caused in the eruptions.

e Suggest one way in which the method used in question d could be improved.

f Identify one observation about major volcanic eruptions compared with major earthquakes.

Topic 10 Hot arid and semi-arid environments

1 Desert adaptations

The table below shows ways in which plants and animals adapt to drought conditions.

Table 10.1 Plants' and animals' adaptations to drought conditions

Plants	Animals
Drought-escaping	
Ephemerals that grow in moist seasons and live through the dry seasons in the seed stage	Animals that enter arid lands only when moisture is available – largely mobile insects, birds and mammals
Drought-evading	
Plants that make economical use of limited soil-moisture supply through wide spacing and reduced leaf and stem surface	Nocturnal, burrowing animals that do not need access to water for temperature control
Drought-resisting	
Succulents that store water and are able to continue growth when soil moisture is not available	Animals that resist drought through physiological processes in which they are able to concentrate their urine, lose little water in faeces, stop perspiration, endure dehydration and still remain active – the camel is a very good example
Drought-enduring	
Drought-dormant plants that aestivate when drought occurs and continue growth when moisture is available	Animals that aestivate and any invertebrates that recover after desiccation

a Define the term *ephemeral*.

b Briefly explain how ephemeral plants cope with drought conditions.

c Explain how animals can be ephemeral.

d Explain how plants can 'evade' droughts.

e Explain how animals may evade drought.

f Suggest a definition of the term *aestivation*.

2 Extreme rainfall in deserts

The table below shows extreme precipitation events in African deserts.

Table 10.2 Precipitation in African deserts

Location	Maximum annual precipitation (mm)	Minimum annual precipitation (mm)	Maximum rainfall in 24 hours (mm)
Tamanrasset, Algeria	159	6	48
Cairo, Egypt	63	3	44
Gadamis, Libya	79	6	17
Dongola, Sudan	60	0	36
Alexander Bay, South Africa	95	22	39
Swakopmund, Namibia	29	0	18

a Choose an appropriate technique to show the maximum and the minimum annual precipitation for the six stations.

b Describe the variations in maximum annual precipitation and minimum annual precipitation that you have drawn.

c Add the maximum rainfall in 24 hours for each of the stations to your graph.

d Describe the relationship between the maximum annual rainfall and the maximum rainfall in 24 hours.

e Compare the data in this table for Cairo with the data for Cairo in Table 10.3 on page 312 of the textbook.

3 Index of aridity

Figure 10.2 on page 309 of the textbook shows the index of aridity.

a Identify the technique used to show the index of aridity.

b State one advantage of the technique.

c Calculate the proportion of the world's land that is arid (semi-arid, arid and extremely arid).

d Explain how the index of aridity is calculated.

e Study Figure 10.1 on page 308 of the textbook, which shows the global distribution of arid areas. Describe the distribution of extremely arid areas.

4 Vegetation cover, runoff and soil erosion

Figure 10.10 on page 318 of the textbook shows the impact of vegetation on runoff and soil erosion.

a Identify the techniques used in the diagram.

b State one improvement that could be made to Figure 10.10.

c Describe the relationship between vegetation type and runoff/soil erosion.

d Briefly explain the changes in runoff and erosion that you have identified in question **c**.

Paper 4 Advanced Human Geography Options

Topic 11 Production, location and change

1 Understanding industrial systems

a In the space below, draw a simple diagram to show the three components of an industrial system.

b How could you 'expand' your diagram to show more detail in terms of the three components referred to in question a?

c Discuss the relative merits and limitations of each type of diagram (simple and detailed).

d Some systems diagrams include 'feedback'. What is feedback and why might it be included on a systems diagram?

2 Interpreting an agricultural systems diagram

Look at Figure 11.12 on page 342 of the *Cambridge International A and AS Level Geography* textbook (second edition), which illustrates an agricultural system.

a Define an *agricultural system*.

b Explain one economic input identified in Figure 11.12.

c Explain one physical input identified in Figure 11.12.

d Discuss the type of decision-making processes that individual farmers have to make.

e Figure 11.12 does not indicate the physical processes that take place on farms. State two such physical processes.

f What are the main outputs of agricultural systems?

g How does the level of output affect the profitability of a farm?

3 Interpreting an industrial location diagram

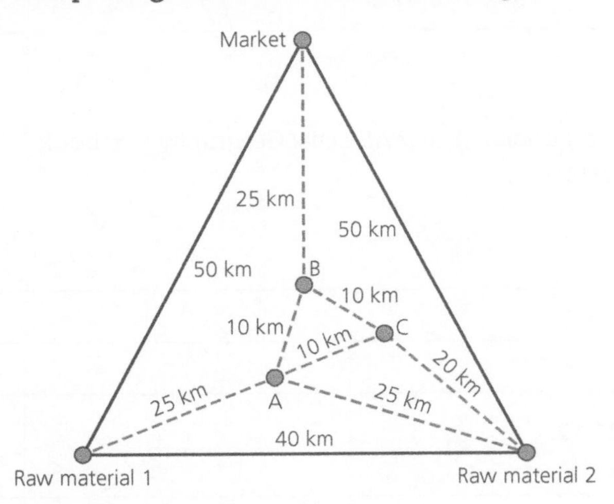

Figure 11.1 Weight-loss location triangle

Figure 11.1 is an adaptation of Figure 11.29 ('Weight-loss diagram') on page 356 of the textbook. Figure 11.1 shows:

- the location of the two raw materials needed to manufacture the end product
- the only market where the product is sold
- the three possible locations for the factory: A, B and C
- the road network and the distance in kilometres between road junctions.

Two tonnes of raw material 1 and one tonne of raw material 2 are needed to make one tonne of the end product. Transport costs are £1 per tonne per kilometre.

a What do you understand by the term *weight loss*?

b Give an example of an industry with a high weight loss.

c What are the units of transport costs used in the example above?

d Which of the three locations, A, B or C, is nearest to the market?

e Complete Table 11.1 below to find the total cost of transport for each location. Which location has the lowest transport costs?

____

Table 11.1 Location triangle cost table

	Location A	Location B	Location C
Cost of transporting 2 tonnes of raw material 1 to factory			
Cost of transporting 1 tonne of raw material 2 to factory			
Cost of transporting 1 tonne of the end product on to market			
Total			

f What are the merits and limitations of the weight-loss triangle in explaining industrial location?

____

____

____

____

____

____

4 Divided proportional circles illustrating changing global oil reserves

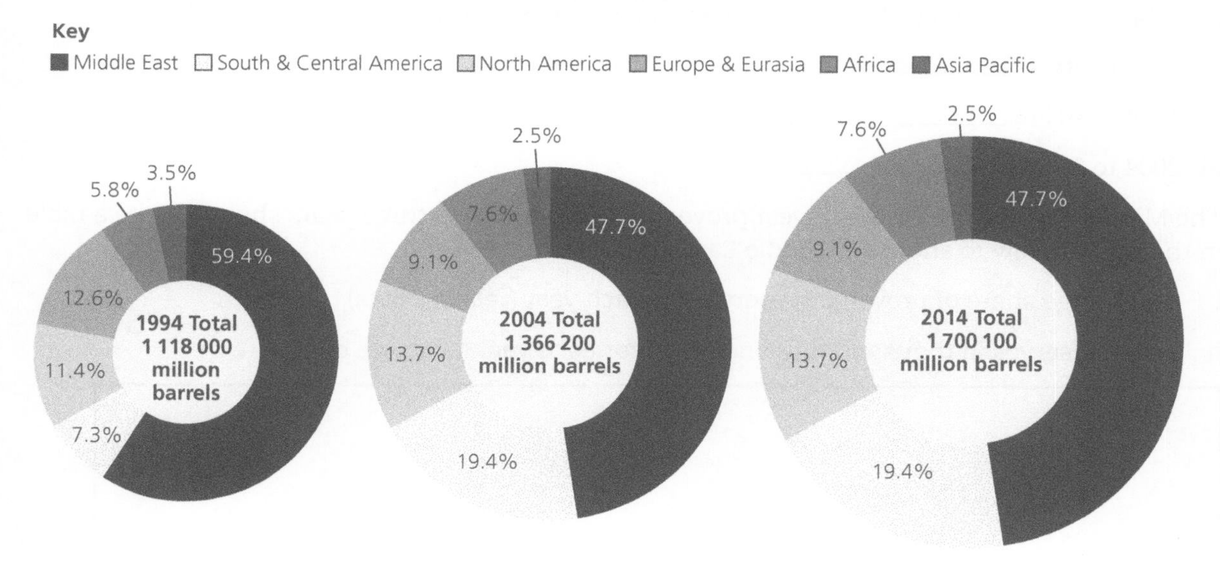

Figure 11.2 Distribution of proven oil reserves, 1994, 2004 and 2014

a Which main graphical technique has been used to show the change in proven oil reserves between 1994, 2004 and 2014 in Figure 11.2?

____

b Explain the construction of the three circles illustrated.

c Which other graphical technique has been used to show the breakdown of proven oil reserves for each of the three individual years?

d Which of the three years has set the order of sequence for the world regions illustrated by the three circles?

e Which region had the greatest increase in the share of global proven reserves between:

i 1994 and 2004? ______

ii 1994 and 2014? ______

iii 2004 and 2014? ______

f By what percentage did global proven oil reserves increase from:

i 1994 to 2004? ______

ii 1994 to 2014? ______

iii 2004 to 2014? ______

g The Middle East has by far the largest proven oil reserves for all three years shown. Write a table in the space below to show the Middle East's:

i percentage share of proven oil reserves for each year (relative data)

ii actual reserves in thousand million barrels for each year (absolute data).

h Describe a data set for which the main technique used for this question might be unsuitable. Justify your view.

Topic 12 Environmental management

1 Diagrammatic representations of environmental degradation

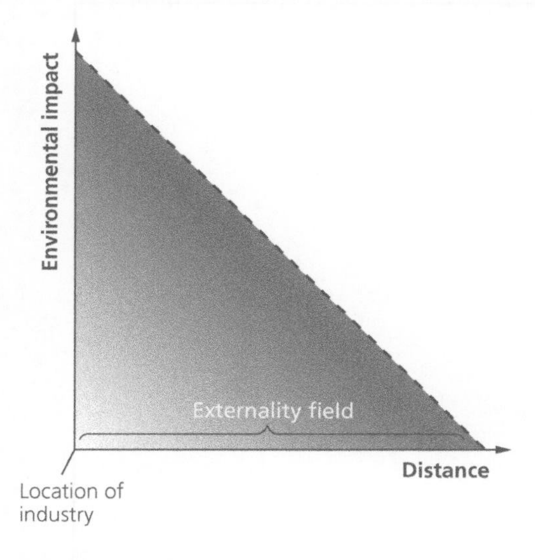

Figure 12.1 Externality field and gradient diagram

Look at Figure 12.1, which shows how the environmental impact of an industrial area changes with distance from the location of the industry.

a Add the labels below to the diagram:

i externality gradient

ii point of maximum environmental impact

iii geographical extent of impact.

b What are *externalities?*

c Explain the terms *externality field* and *externality gradient.*

d Show on the diagram what you would expect to happen if additional polluting factories were sited at the 'location of industry'. Explain what you have drawn on the diagram.

e The environmental impact is generally the major negative externality of a large industrial area. What other externalities might there be, both negative and positive?

f In the space below, draw and label an environmental Kuznets curve to show how the rate of environmental degradation changes with economic development.

g Explain the curve you have drawn.

Topic 13 Global interdependence

1 Illustrating trade statistics

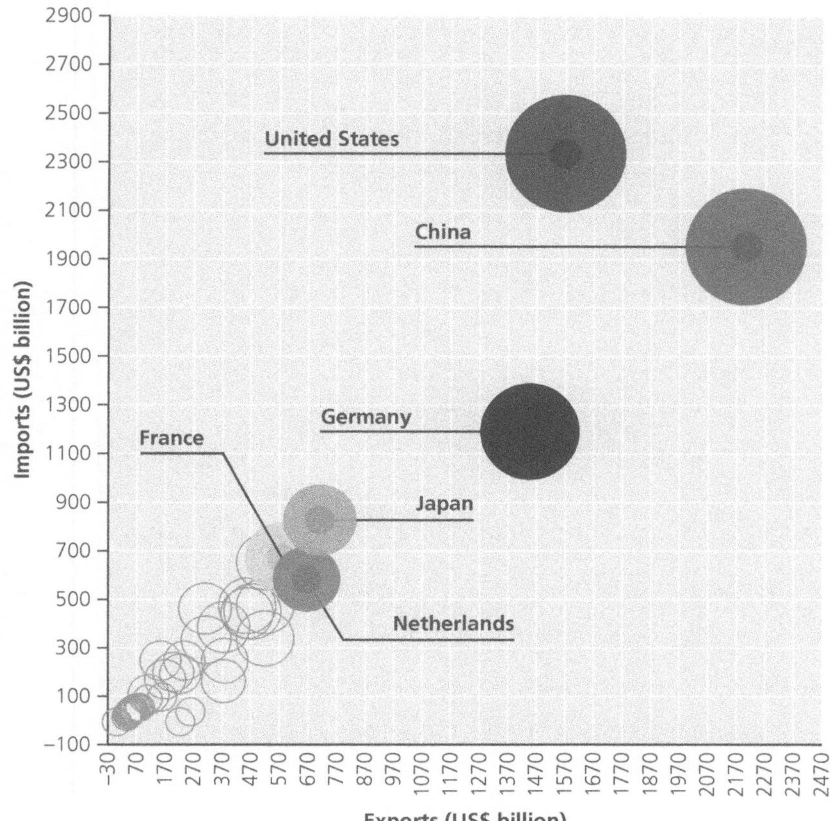

Figure 13.1 Scatter graph with proportional circles

a What is meant by the term *merchandise trade?*

b Briefly describe the scales used in Figure 13.1.

c Look at the six nations named in Figure 13.1. Account for the different sizes of the circles representing these nations.

d The centre of each circle represents the balance between each country's exports and imports. Write a table showing these figures (approximately) in the space below.

e Which countries have a merchandise trade surplus?

f Which countries have a merchandise trade deficit?

g What important aspect of trade is not considered in Figure 13.1?

2 Interpreting diagrams that illustrate the economic impact of tourism

Look at Figures 13.48 and 13.49 on page 441 of the textbook. The former illustrates the concept of economic leakage, while the latter outlines the multiplier effect of tourism.

a Tourism brings valuable foreign currency to many LICs. However, its value is often overrated because of a number of factors, in particular economic leakages. What are economic leakages?

b Explain two of the economic leakages shown in Figure 13.48.

c What is the estimated rate of economic leakages in tourism from LICs to HICs?

d Explain the statement 'tourism is a labour-intensive industry'.

e Why is tourism not of greater benefit to LICs in terms of employment?

f The *multiplier effect* is an important term frequently used in economics and geography to explain positive economic change in a country or region. Define the *multiplier effect*.

g Briefly explain how the multiplier effect can operate in the tourism industry.

Topic 14 Economic transition

1 Photographic evidence of development

Figure 14.1 Bridge in a chronic state of repair in southern Mongolia

Look at Figure 14.1 above and also at Figure 14.8 on page 454 of the textbook, which shows an open-pit toilet.

a What type of photographs are these images?

____

b Describe what the photographs show in terms of the level of development in this country.

____

c Suggest and justify photographs of two other aspects of development that would help provide a fuller picture of the level of development in this country.

____

d What are the merits and limitations of photographs such as those referred to above as a source of evidence regarding economic development?

____

e How might aerial and satellite photographs add to the study of development?

2 **Techniques to illustrate employment structure and changes over time**

a Name a graphical technique that can be used to illustrate the distribution of employment between the primary, secondary and tertiary sectors.

b Explain how this graphical technique is constructed and how it can be interpreted.

c Suggest two other areas of geographical analysis for which this graphical technique can be used.

d Briefly consider the merits and limitations of this technique.

e In the space below, draw a diagram to show how employment structure generally changes over time. Provide a title and label the axes of the diagram clearly. Also label all the lines you have drawn on the diagram.

f How could you use this diagram in conjunction with the graph you named for question **a**?

3 Economic indicators of development

a In terms of the level of economic development in a country, what does the abbreviation GDP stand for?

b Explain how this measure is calculated.

c How is GDP per person calculated and why is GDP per person data more often used than GDP data alone?

d Briefly state the merits and limitations of this measure of development.

e Increasingly, GDP or GNI data is shown at purchasing power parity ($PPP). Explain what this means and why it is important.

4 The Human Development Index and ranking systems

a When was the Human Development Index introduced and what is its purpose?

b Draw a diagram in the space below to explain how the Human Development Index is calculated for an individual country.

c Look at Figure 14.12 on page 455 of the textbook, which shows a map of the HDI. What are the four classes of human development shown on the map?

d Describe the locations of the countries with the best and worst levels of human development.

e If you were to add one more indicator of development to the Human Development Index, what would it be? Justify your choice.

f What is the objective of producing a list of countries in rank order such as in the Human Development Index?

g Give one limitation of a basic ranking system.

5 Assessing and illustrating income inequality within countries

a Name a technique that can be used to show the level of income inequality in a country.

b How are the figures for individual countries calculated?

c Name one world region where income inequality is high and one world region where income inequality is relatively low.

d Suggest how you might illustrate the Gini coefficient for all countries on a world map.

e The Lorenz curve is a graphical technique that can be used to illustrate income inequality within a country and to show how this phenomenon can change over time. Draw a fully annotated diagram of a Lorenz curve on the grid below.

Cambridge
International AS and A Level

Geography
WORKBOOK

Perfect for use throughout the course, this workbook supports students using the Cambridge International AS and A Level Geography textbook. It provides practice for the skills which will be assessed in the examination and additional support for students to develop their geographical skills.

- Includes examples of different types of graphs, charts, maps and diagrams, with explanations
- Ensures students are familiar with a variety of source materials
- Suggested answers are available on the accompanying Teacher's CD

www.hoddereducation.com

ISBN 978-1-471-87376-8